CAMBRIDGE COUNTY GEOGRAPHIES

General Editor: F. H. H. GUILLEMARD, M.A., M.D.

MERIONETHSHIRE

Cambridge County Geographies

MERIONETHSHIRE

by

A. MORRIS, F. R. Hist. Soc.

With Maps, Diagrams and Illustrations

Cambridge :
at the University Press
1913

CAMBRIDGE UNIVERSITY PRESS
Cambridge, New York, Melbourne, Madrid, Cape Town,
Singapore, São Paulo, Delhi, Mexico City

Cambridge University Press
The Edinburgh Building, Cambridge CB2 8RU, UK

Published in the United States of America by Cambridge University Press, New York

www.cambridge.org
Information on this title: www.cambridge.org/9781107629288

First published 1913
First paperback edition 2013

A catalogue record for this publication is available from the British Library

ISBN 978-1-107-62928-8 Paperback

PREFACE

THE author desires to acknowledge his indebtedness to various works in English and Welsh on the history and antiquities of Merionethshire, especially the articles of the late W. W. E. Wynne of Peniarth on the architecture of the most remarkable of the churches. His thanks are due to Mr Pryce Williams of Towyn for assistance rendered in the chapter on *Fisheries and Fishing Stations*, and to Mr D. A. Jones of Harlech and the late Thomas Ruddy of the Palé Gardens for help in the preparation of the chapter on *Natural History*.

A. MORRIS.

October 1913.

CONTENTS

ILLUSTRATIONS

MAPS

The illustrations on pp. 5, 7, 9, 13, 14, 16, 20, 21, 22, 24, 26, 33, 45, 47, 59, 72, 73, 75, 85, 110, 111, 119, 128, 129, 133, 137, 144 and 157 are from photographs by Messrs F. Frith and Co.; those on pp. 61 and 103 from photographs by Mr D. H. Parry, Harlech; those on pp. 121 and 151 from photographs by Messrs George and Son, Corris; those on pp. 112 and 114 from photographs by Mr W. M. Dodson, Bettws-y-Coed; that on p. 94 from a photograph by Dr Guillemard; that on p. 12 from a print published by Mr R. L. Jones, Machynlleth; that on p. 115 from a photograph by Mr Jones, Dolgelly; that on p. 124 from a photograph by Mr Arnfield, Dolgelly; that on p. 146 from a photograph kindly supplied by Mrs Ellis; those on pp. 101 and 102 are reproduced from *Archaeologia Cambrensis* and the *Archaeological Journal* respectively; the sketch map facing p. 116 is from a drawing by Mr C. J. Evans.

1. County and Shire. The name *Merionethshire*. Its Origin and Meaning.

The division of Wales into shires first took place in the reign of Edward the First. Before the conquest of Wales by that monarch there was no division of the Principality into shire ground as understood in English annals. The *Shire* (i.e. the part *shorn* off, or cut off, from the Anglo-Saxon word *scir*) was a Saxon institution brought into use at an early period, as early as the seventh century. In the code of laws of Ina of Wessex, we find portions of the country under his rule divided into *scir* ground, and each division was placed under an officer who was styled a *scir-gerefa*, i.e. a shire-reeve or sheriff. He was the natural leader of the shire in war and peace. His duties were to look after the king's rights, dues and fines, and he acted as the sovereign's representative as regards finance and the execution of justice.

County is a word of Norman origin (*comté*) which came into use in our country after the Conquest, when the administration of each shire was entrusted to a great earl or baron, who was often a count (*comte*), i.e. a companion of the king.

The shiring of Wales was the direct outcome of the extension of English influence into our land. It took place upon two separate occasions, the first as stated above, and the second in the reign of Henry the Eighth. Consequently the shires of Wales do not stand in the same relation to the early history of the particular districts of which they are a share, as the real shires of England proper stand to old English history. They are really administrative districts formed for convenience, rather than organic divisions of land and people like Sussex and Kent, which correspond to original tribal kingdoms. Of the Welsh counties Anglesey's insular position gave it a unity and compactness of its own, but as regards the others, Cardiganshire alone in extent of territory and distinctive characteristics is in an analogous position to that of Sussex and Kent among English counties. It probably corresponded with the ancient principality of Ceredigion, and to this, perhaps, the strong local feeling and distinctive type of character still associated with that county are due. The other counties have, however, been built up of the immemorial territorial divisions (hundreds and commotes) of the Cymry.

The county of Merioneth is one of the eight counties which came into existence by the Statute of Rhuddlan in 1284. The name, however, is of much earlier date as the name of a cantrev. In its Welsh form of Meirionydd we are taken back to a period some eight centuries earlier.

The tradition is that about 420 A.D. Cunedda, a powerful British chief who held his court at Carlisle, was invited by his kindred, the Brythons, to come and assist them, as they were sore pressed by the Gwyddyl or Goidels from across the Irish Sea. In right of his mother, as we are told in the Welsh pedigrees, Cunedda was able to claim large tracts of territory in Wales. He therefore most readily responded to the appeal, and by the aid of his numerous sons succeeded in expelling the Goidels from the greater part of the territory. Cunedda's men, it is recorded, settled permanently in the land, and so did his sons, except the eldest, named Tybiawn, who had died some time before in Manaw Gododin, as the territory of the north was called.

The names of the sons have survived in the territories which they wrested from the Gwyddyl. Ceredig occupied Ceredigion (Cardiganshire); Arwystl seized upon Arwystli, a part of Montgomeryshire; Edeyrn made his abode in Edeyrnion in our present county; Einion possessed himself of Caereinion in Montgomeryshire. The sons of Tybiawn were likewise granted their shares, in right of the eldest son. Maelor obtained Dyffryn Maelor, and Meirion possessed the territory called Cantrev Meirion, "the Hundred of Meirion," which in its turn gave its name to Meirionydd, and the county of Merioneth.

By the Statute of Rhuddlan there were added to the cantrev of Meirionydd the commotes of Penllyn, Edeyrnion, and Ardudwy, and these together constituted the shire of Merioneth until the time of Henry VIII. When the Principality became ripe for its union with England in the time of the Welsh sovereigns, the Tudors, an "Act of Union" was passed, by which five new shires

were created from the Marcher lordships. This Act
added to the county of Merioneth the lawless lordship of
Mawddwy.

2. General Characteristics.

Merionethshire is a maritime county of North Wales,
washed on its western side by Cardigan Bay, and bor-
dered on the north, east, and south by the counties of
Carnarvon, Denbigh, Montgomery, and Cardigan respectively.

It is more mountainous than any of the North Wales
counties with the exception perhaps of Carnarvonshire.
Its deep and secluded valleys, with the ruggedness and
variety of its elevated districts, give it a particular charm
and interest. The varied panoramic views from its
heights surpass anything to be seen in Wales.

Portions of the county, by the nature of its rocks, are
devoted to the industry of slate-quarrying. The best
slate in the world for roofing purposes is worked in
various parts of the county, but mainly in the north.
Ours, too, is the only county in Wales in which gold has
been found in quantities sufficient to pay for working;
but, in the main, Merionethshire is an agricultural and
pastoral county, the great proportion of the people being
devoted to husbandry.

Merionethshire is one of the most Welsh in customs
and habits of all the counties of Wales. Its people have
not been influenced to the same extent as other Welsh

The Vale of Festiniog

counties by the influx of the English wave. In the most numerous instances business and trade dealings are carried on in the vernacular, and the native inhabitants treasure their ancient language as the worthiest of their inheritances.

The county has figured largely in the history of the Principality from the earliest times. Its remains of the prehistoric past form an interesting chapter in the story of our land. The Brythonic wave of our Celtic forebears pushed itself from the plains of England into Wales by way of large tracts in this county, and terminated like the point of a broad wedge at the mouth of the Dovey. From this fact has arisen the name of the Brythonic tribe in the second wave of Celtic migration, *Yr Ardyfiaid*, or as it was known to the Romans, the *Ordovices*. The Goidels and the Brythons have left traces of their occupation in the vast number of tumuli, menhirs, stone circles, and cromlechs now seen in elevated situations in various parts of the county.

The remains of Roman times are also very interesting. These comprise military roads, camps, and stations in all parts of the county. Other remains, such as coins, inscribed stones, and Samian ware prove that Roman civilisation held sway in secluded corners of this county as well as in the more accessible parts of England and the borders, while the ruins of castles and ecclesiastical buildings show it to have been a not unimportant territory in medieval times.

Its rivers are famous all the world over for their incomparable scenery. The Dee, with its great inland sheet of water snugly sheltered by the Arenigs and the Berwyns, has been more sung about and visited than most of the rivers of Cambria. The Mawddach with its broad tidal estuary and its numerous rushing contributory streams has noble scenery to show, and the district through which it flows is sometimes called the British Tyrol. Its lakes and waterfalls are equal in beauty to those in any part of

The Mawddach, from Panorama Walk

the kingdom, and are an unceasing source of attraction to hosts of sight-seers at all seasons of the year. The woody character of its valleys and uplands make it a delightful land. No part of its surface can be said to be tame or monotonous. From every standpoint our county of Merioneth is one of the most charming and interesting of all the Welsh counties.

3. Size. Shape. Boundaries.

Merionethshire is one of the largest counties of North Wales, comprising an area of 602 square miles with a superficial surface in the administrative county of 418,475 acres excluding water. Its water area totals 3897 acres. It occupies nearly one-twelfth of the whole area of Wales, and ranks as seventh in point of size of the twelve Welsh counties. In the geographical or ancient county area it would take rank as sixth in Wales.

Its extreme length, from north-east to south-west, measured in a straight line drawn along the southern contour of the county from Berwyn on the Dee to Aberdovey on the Dovey estuary, is 46 miles. Its greatest breadth, measured from Llyn-y-Ddinas in the north, near the village of Beddgelert, to near Mallwyd on the borders of Montgomeryshire, is 29 miles.

In shape, speaking generally, the county has the appearance of a scalene triangle, having its shorter side on the west where it faces Cardigan Bay. The base, its longest side, lies contiguous to Montgomeryshire for a length of 37 miles, with the remaining nine miles touching Denbighshire. The apex of this triangle is at the west corner of Llyn-y-Ddinas, whilst the angles of the base are respectively at the village of Berwyn on the Dee and at Aberdovey.

In a perambulation of the limits of the county it will be well to make our start at the apex of this irregular triangle. We shall be compelled to observe that, with the

exception of the west side, our county is so circumscribed
by high mountains that there are only a very few artificial
boundaries necessary.   Nature has fulfilled her part in
an admirable manner ; she has supplied the county with
natural boundaries in her high mountains, rivers, and
sea.

Leaving Llyn-y-Ddinas the boundary line takes us
first to the top of the Glyders, high mountains forming

Aberdovey

an offshoot of the Snowdon group.   Hence we proceed
by an arc of a circle to the north of the steep and rugged
Cynicht, until it encloses the slate district of Festiniog
within its bounds.   The circumference of this arc of a
circle descends by Llyn-y-Dywarchen, a charming sheet
of water, and leaves the limits of Carnarvonshire to enter
those of Denbighshire.

We now follow it a little to the north of the mountain called Arenig Fach, from which it proceeds in a north-easterly direction across the elevated expanse of the Gylchedd, to ascend the ridge of Carnedd Filast and then drop into the valley of the upper course of the Alwen. It follows this little stream for about two miles until it approaches Cerrig-y-Drudion on the Denbighshire side of the boundary. The line of demarcation now takes a southerly course, and forms what may be called three-fourths of a circle to cross the Alwen again about two miles to the south of Bettws Gwerfil Goch. It assumes a northward direction a little to the west of this village, and reaches Llanfihangel Glyn Myfyr. Here it proceeds eastward for four miles, and then by an artificial limit makes for the valley of the Clwyd, which it crosses at the village of Derwen. A mile beyond this north-eastern limit an artificial boundary again marks the line of demarcation on the eastern side until it arrives at the village of Berwyn on the Dee.

From Berwyn our direction is now south-west by a zigzag course until we reach the summit of Moel Ferna in the Berwyn group. Proceeding along the length of this chain of mountains for nearly ten miles we arrive at Cader Berwyn, and here we leave Denbighshire to beat the bounds of Montgomeryshire. The boundary line continues along the Berwyn chain as far as the pass of Bwlch-y-Groes. On our right is the valley of the Dee with Bala Lake in its course; and on our left, in closer proximity to the mountains, is Lake Vyrnwy, the great artificial reservoir of the city of Liverpool. These mountains separate the basin of the Dee from that of the Vyrnwy and the Severn.

At Bwlch-y-Groes we cross the remarkable pass connecting the two valleys. The scenery here is very wild and picturesque. We proceed for a few miles to the south and cross the summit of Carreg-y-bîg, leaving Llan-y-Mawddwy nestling in the valley below. Five miles further to the south we arrive at Nant Dugoed, where we cross the turnpike road from Llanfair Caereinion to Mallwyd, which is only five miles distant. The line of demarcation, once more an artificial one, passes to the other side of the valley of the upper Dovey; this river constituting the boundary between Mallwyd and the village of Aberangell. It then crosses the ridge separating the valley of the Dovey from that of its tributary, the Dulas. This latter stream in its course by Corris to Machynlleth, where it joins the Dovey, forms the boundary between our county and Montgomeryshire.

From Machynlleth to the sea the Dovey is again the dividing line until we almost reach the estuary, when it leaves Montgomeryshire and has Cardiganshire as its neighbouring county to the south.

On the west, from the estuary of the Dovey to the mouth of the Glaslyn river on the borders of Carnarvonshire, Merionethshire is washed by the sea. For the remainder of the western limits of the county the Glaslyn is the dividing line. It comes from Llyn-y-Ddinas and proceeds by Beddgelert through the gorge of Aberglaslyn, then by a sinuous course across the reclaimed Morfa it finds its way to the sea at Traeth Mawr.

By this perambulation of the county across its rugged and broken peaks it will have been realised how difficult it would be to find a district so hemmed in by mountains as our county of Merioneth. This circumscribed character of the land has made for the striking and distinguishing characteristics of its people, and this influence must have been exerted with tenfold force in the days before railways and excellent roads opened out the interior.

Rhaiadr Cwm, near Festiniog

(*From an old print by W. Radclyffe, after a painting by David Cox*)

4. Surface and General Features.

As we have seen in our last chapter Merionethshire is an exceedingly mountainous county. In some respects it may be said that of all the Welsh counties it is the most diversified, for mountains and hills occur more universally than in the other counties. The only lowland territory apart from its deep valleys is the narrow strip along the coast between the Ardudwy mountains and the sea. Its elevations are not as high as those of the neighbouring county of Carnarvon, but many of its peaks are very fine. They abound in bare precipitous cliffs and rugged heights and the slopes of many of them are streams of sliding fragments of wrecks of stone.

Cader Idris: the Summit from the Saddle

No isolated and solitary peak of the Welsh mountains shows such a wreck of stone as Cader Idris. With an elevation of 2927 feet above the sea-level, it marks the starting point from which a long chain of primitive mountains extends in a north-east direction to the Berwyns

and on to the borders of Shropshire. This chain has a
fine array of towering heights. The Aran Mawddwy is
higher than the Gader, and reaches an altitude of
2970 feet; and there are others, such as Aran Benllyn,
2901 feet, which closely approximate the Gader.

Cader Idris throws off spurs to the south-west which

The Bird Rock, Towyn

gradually decline in elevation the further we proceed,
until we arrive at the estuary of the Dovey. A feature
of one of these ridges is the Craig Aderyn—the curious
"bird-rock" as it is called, which is the home of countless
sea-fowl. It lifts its bold and isolated head some six or
seven hundred feet above the banks of the Dysynni river.

A little to the north-east on the slope of another hill of this ridge are the ruined remains of Castell-y-Bere, a famous old medieval fortress, while on the opposite hill across the valley of the Dysynni is the reputed cave refuge of the great Cymric hero, Owain Glyndwr.

Proceeding southwards in the direction of Aberdovey we have Trum-tair-taren, Moel-y-Geifr, and Trum Gelli, which run on to the ridge known as Mynydd Bychan ; and still further south we come to Mynydd-y-Llyn, having Llyn Barfog, a charming little lake, nestling at its base.

The impressive Berwyn chain, occupying the south-east border of the county and forming the southern watershed of the river Dee, forms a remarkable contrast to the bareness of the Gader group by the richness of its vegetation, especially on the Deeside slopes. Some of its heights very nearly approximate the altitude of Cader Idris and the two Arans. The chief are Cader Fronwen 2575 feet, Cader Berwyn 2716 feet, Moel Ferna 2070 feet, Moel-y-Geifr 2055 feet, and Moel-y-Cerrig-duon 2050 feet.

Nine miles to the north-east of Dolgelly the range of the Gader throws off a spur to the north to join the Arenigs. This spur forms the dividing watershed of the Dee and the Mawddach. The country here is wild and secluded, and has remains of *carneddau* and tumuli in abundance.

The Arenig series of mountains occupies the whole of the north of the county. The most remarkable heights are the Arenig Fawr with its double-headed ridge, 2800 feet, Moelwyn Mawr 2527 feet, Cynicht 2763 feet, Arenig Fach 2250 feet, and Carnedd Filast 2197 feet.

Extending from the valley of the Mawddach to that of Maentwrog, and running nearly parallel to the coast, we have the interesting group known as the Llechwedd chain, or mountains of Ardudwy, which run up from Barmouth and terminate in the Diphwys at a height of

Barmouth: Diphwys

2467 feet. Beyond this rises Craig-y-Cau 2063 feet, the Llethar 2475 feet, Rhinog Fach 2300 feet, and Rhinog Fawr 2362 feet.

The famous Roman road known as the *Via Occidentalis* or the Sarn Helen traversed these mountains. Between the Rhinog Fawr and Llyn Cwm Bychan there are what are usually called the Roman Steps. It is surmised that these steps were made by the Roman soldiers to facilitate the conveyance of the ores from the mines. The sides of these mountains seem to have been rent asunder by some mighty convulsion into a thousand precipices, forming at their tops rows of shelves which the native-folk compare to sills of dovecotes, and call *Cerrig Colomenod*, i.e. "Rocks of the doves." The scenery of this extraordinary pass and that of Bwlch Drws Ardudwy, for its wild character and its ruins of stone, rivals in its way anything to be seen in the Alps, and the views from the summits of the mountains are magnificent. The panoramic view overlooking Barmouth and the Mawddach estuary, and the "Precipice Walk" of the Upper Mawddach gorge on the slopes of Moel Cynwch, near the ancient domain of Nannau, are especially striking. Moel Offrwm, "the Mountain of Sacrifice," is close by, and to the north of Llanfachreth in the same district stands Rhobell Fawr, 2409 feet, a solitary eminence.

5. Watershed. Rivers.

The chief watershed of the county is the high ridge dividing Merionethshire into east and west, referred to in the previous chapter. The fall of the east drains into the Dee and its tributaries, whilst that of the west forms the streams which flow into Cardigan Bay. The most important rivers are the Dee, the Mawddach, the Dovey, the Dysynni, the Dwyryd, and the Glaslyn.

The Dee, the principal river of North Wales, rises by a small streamlet in the Dduallt at an elevation of

2000 feet above the sea-level, and about four miles to the west of Bala Lake. The Great Western railroad from Dolgelly, after its ascent to Drws-y-nant near Aran Benllyn on the highest ridge of the wild watershed, follows in its descent to Llanuwchllyn the course of the infant Dyfrdwy, as the Dee is called in the vernacular. The Dyfrdwy before entering the lake receives on the right the Twrch from Aran Benllyn, and on the left the Lliw from Moel Llyfnant, which washes the base of the slope where the remains of Carn Dochan Castle stand.

The three streams meet at the little village of Llanuwchllyn, and the united waters enter Bala Lake, or Llyn Tegid, the largest natural lake in the whole of Wales. At the eastern end of the lake stands the little town of Bala, and here the Dee leaves the lake and receives the Tryweryn, a tributary rising in the Arenig Fach, which rushes down a strong clear stream, through charming scenes, to pass through the wooded gorges of Rhiwlas with its fine old mansion. The waters of the Tryweryn and Dee unite in the flat meadows below the lake.

The Dee now wends its way for twelve miles through the sweet Vale of Edeyrnion, a broad and noble river. In times of heavy rain it is a sheet of seething foam rushing over beds of broken boulders, but in summer days a placid waterway like a broad band of silver. Its course leads through the woody glades of the villages of Llandderfel and Llandrillo. On our right we pass the famous battle-ground of Crogen on the summit above, and on our left is the mouth of the Alwen, a tributary stream from the

Denbighshire border. We now come to the well-wooded enclosures of Rûg, where in the twelfth century, Gruffydd ap Cynan, King of Gwynedd, was entrapped, and to Corwen, an old-world market town, tucked under the dark shoulders of the mountains. Caer Drewyn is on the left bank, a famous encampment in ancient times. After leaving Llansantffraid Glyndyfrdwy the river passes the village of Berwyn, and there leaves the county to continue its course through Denbighshire.

Between Llansantffraid bridge and Llangollen in the month of April, when the spring trout-fishing is at its best, we may see the old-world coracle used for fishing. The Dee is the only river in North Wales where this survival of the ancient Britons is still put to practical and common use.

The Mawddach has its rise in the same central watershed as the Dee. It comes from the spur known as Craig-y-Ddinas, and then descends through Cwm Allt-lwyd into one of the most lovely glens of picturesque Wales. Before receiving the Afon Gain, which flows from the upper reaches of the central watershed, the main stream has passed the Gwynfynydd gold mine, which yielded considerable quantities of the valuable ore in years gone by. The famous Pistyll-y-Cain waterfall is at this spot. Here also may be seen Rhaiadr-y-Mawddach, and about a mile above Ty'n-y-Groes another cataract called Y Rhaiadr Ddu, situated near the confluence of the Camlan with the Mawddach.

The Mawddach is joined at the Ganllwyd by the Afon Eden from Craig Ddrwg in the Ardudwy mountains. The

main stream then takes a straight course to Llanelltyd,
and passes the ruins of Cymmer Abbey on its left bank,
before it receives the Afon Wnion from Drws-y-Nant
Uchaf. The course of the Wnion for the greater part
of its length to Dolgelly is in a deep chasm of serrated

Prysor Valley, Rhaiadr Ddu

rocks. It receives the Clywedog two miles above Dolgelly, which is famed for the rushing cataracts of the
"Torrent Walk."

At Llanelltyd the Mawddach becomes a tidal river,
the tide flowing up to the bridge near the village. Upon
its approach to Penmaenpool the river gradually widens

into a broad waterway.   At Bontddu on the right bank
is the Clogau gold mine, which is systematically worked
at the present time.

The mountains of Ardudwy are drained by the Afon
Artro which rises in Llyn Cwmbychan near the Roman
Steps.   It flows swiftly over a rocky bed, hidden at times
by dense woods, until it arrives at Llanbedr.   Before

The River Artro at Llanbedr

reaching the latter village the Artro receives the Nant
Col from the gorge of Bwlch Drws Ardudwy.   The
united waters find their way to the sea at the northern
end of Mochras Island.

The Dwyryd, which enters the sea at the Traeth
Bach, having flowed through the beautiful vale of Festiniog, is formed, as the name implies, by the union of two

streams. These are the Goidol and Tegwel, which first unite and are afterwards joined by the Cynfael. The Goidol drains Llyn Cwmorthin to the north of Tany-grisiau, whilst the Tegwel comes from Carreg-y-Fran to the east of Llyn-y-Manod. Before these streams unite at Rhyd-y-Sarn, the Tegwel has passed by Beddau Gwyr Ardudwy—"the graves of the men of Ardudwy"—with

Llanfihangel-y-Pennant

Llyn-y-Morwynion a little to the south. A little below Rhyd-y-Sarn the Cynfael flows in, and from the con-fluence the united streams are called the Dwyryd. In the bed of the river here arises the singular isolated column of rock known as " Hugh Llwyd's pulpit." The Dwyryd continues its course through a beautiful valley by the little village of Maentwrog. About a mile below Maentwrog it receives the Afon Prysor, which rises in the Graig Wen, an elevation to the east of the Roman station of Tomen-y-Mur, and flows by Trawsfynydd in a most circuitous course.

The Dovey rises in Craiglyn Dyfi on the eastern declivity of Aran Mawddwy. It flows by Llan-y-Mawddwy, Dinas Mawddwy, and Mallwyd, and is fed by numerous contributary streams, the Dulas, the Llefeni, the Geryst, the Clywedog, and the Pennal, and flows for twelve miles of its course through Montgomeryshire.

The Dysynni rises in the southern declivities of Cader Idris and flows through Llanfihangel-y-Pennant, a small village in a beautiful situation, then by Castell-y-Bere and Craig Aderyn to skirt the shady glades of Peniarth and the little village of Llanegryn. It enters the sea about two miles to the north of Towyn.

6. Lakes.

The lakes of Merionethshire are of exceedingly great interest on account of their situation, their beauty, and the wealth of folk-lore connected with them. First and foremost we have Bala Lake, or Llyn Tegid, lying between the Berwyn chain and the Arenigs. Then come the remarkable series known as the Cader Idris group, surrounding the base of that mountain. Next to these come those of the Ardudwy mountains, small in size but charming in situation. And finally we have those of the Festiniog district, numerous and rich in traditionary lore.

Bala Lake in the valley of the upper waters of the

Dee, the largest sheet of natural fresh water in Wales, is
1084 acres in extent, but it has now been exceeded in
size by the artificially constructed Vyrnwy reservoir on
the other side of the Berwyns, which has an area of 1121
acres. Its length is about three miles, with a breadth in
the widest part of nearly one mile.

The lake, like so many of the Welsh lakes, is not

Bala Lake and Llanycil Church

devoid of a legend as to its origin, though it is too long to
give here. Its Welsh name, Llyn Tegid, takes us back
to a remote past. Tegid Foel is said to have been the
husband of Ceridwen, the traditional mother of Taliesin,
the seer, and his dominion comprised the territory in
which the lake is situated, though according to the legend
it was not in his time that the lake was formed.

The lakes of the Cader Idris group are Talyllyn, Cae, Tri Graienyn, Aran, Gader, Gafr, Gwernan, Wylfa, and Creigenen. The largest of these is the Talyllyn lake at the southern foot of the mountain. It is sometimes known as the Mwyngil, i.e. "the Peaceful Retreat," a very appropriate name for this beautiful and secluded stretch of water, which is about two miles long and half a mile broad. Verdant meadows and sequestered homesteads surround it, whilst the rugged grandeur of Cader Idris towers above. The Dysynni river drains it, having its outlet at the eastern end. Llyn-y-Cae is in a chasm of the mountain above Talyllyn and is best seen from the summit of the Gader. Llyn-y-Tri Graienyn, the "Lake of the Three Pebbles," is situated at the side of the road from Corris to Dolgelly. The pebbles—three huge boulders weighing many tons—according to the legend, were shaken out of the shoe of the giant Idris. On the northeast side is Llyn Aran, drained by the stream of the same name, which flows through the town of Dolgelly to join the Wnion. Llyn-y-Gader lies at the foot of the Fox's Path, which leads to the summit of Cader Idris. It is sheltered by high precipitous cliffs. Within a distance of half a mile is Llyn Gafr, called by this name because of the large herds of goats which grazed its banks in former times. On the side of the road from Dolgelly to Cader Idris is Llyn Gwernan, a beautifully clear lake, but in summer filled with sedge and vegetable growth. Llyn Creigenen lies on the elevated ridge above Arthog. Its waters help to form the beautiful falls of that place.

In the Ganllwyd neighbourhood on the northern side

Talyllyn

of the Mawddach estuary is Llyn Cynwch. This is near the old mansion of Nannau, and is the reservoir supplying Dolgelly with water.

We now turn to the lakes of the Ardudwy country, among which is Llyn Cwmbychan, at the foot of the ridge known as Graig Ddrwg. It is a small though charming sheet of water within a walk of Harlech. To the north are Llyn-yr-Eiddew Mawr and Eiddew Bach ; these are drained by the Artro. Llynau Tecwyn Uchaf and Isaf lie in the mountains between Talsarnau and Trawsfynydd ; surrounding them is a marvellous wreck of stones in which some archaeologists claim to find traces of "a hitherto unknown British town." Coming southwards into the valley of the Ysgethin, and at the foot of the Llawllech chain, we have three lovely lakes named the Bodlyn, Irddyn, and Dulyn. Across the Diphwys is Llyn Cwm Mynach, the water of which is carried down to Bontddu by the Afon Mynach. There are many more of these small lakes in this district, some of which are Llyn Du, Llyn-y-Fedw, Llyn Pryfed, and Llyn Dywarchen.

In the Festiniog district there are several beautiful sheets of water, some of which are of considerable size. The best known are Llyn-y-Morwynion, Y Dywarchen, Y Manod, Bowydd, Conglog, Cwmorthin, Llynlilyn, Llynau-y-Gamallt, Llyn Newydd, Y Garn, Tryweryn, Arenig Fawr, and Arenig Fach. Of these the most famous is Llyn-y-Morwynion—"Maidens' Lake,"—because of the legend connected with it, which in Welsh lake-lore is as famous as that of the Sabine women in classic story.

7. Geology.

The term *rock* in Geology is used without reference to the hardness or compactness of the material. The hardest rock, as well as the softest, crumbles into sand and dust by exposure to the atmosphere, and geologists speak of loose soil, layers of sand, pebble, or clay by the same term as they do of slate, limestone, or granite.

Rocks are divided roughly into two classes, (1) those laid down mostly under water, called *sedimentary* or *aqueous*, (2) the *eruptive* or *igneous*, i.e. those due to fire and volcanic action.

The first kind may be compared to sheets of paper lying one over the other. These sheets are called *beds*, and are usually formed of sand (often containing pebbles), mud or clay, and limestone, or mixtures of these materials. They are laid down as flat or nearly flat sheets, but may afterwards be tilted as the result of movement of the earth's crust, just as we may tilt sheets of paper, folding them into arches and troughs, by pressing them at either end. Again, we may find the tops of the folds so produced worn away as the result of the action of rivers, glaciers, and sea-waves upon them, just as we might cut off the tops of the folds of the paper with a pair of shears.

The eruptive or igneous rocks have been melted under the action of heat and become solid on cooling. When in the molten state they have been poured out at the surface as the lava of volcanoes, or have been forced into other rocks and cooled in the cracks and other places of weakness. Much material is also thrown out of volcanoes as volcanic ash and dust, and is piled up on the sides of the volcano. Such ashy material may be arranged in beds, so that it partakes to some extent of the character of the first of the two great rock groups.

The relations of such beds are of great importance to geologists, for by them we can classify the rocks according to age. If we take two sheets of paper, and lay one on the top of the other on a table, the upper one has been laid down after the other. Similarly with two beds, the upper is also the newer, and the newer will remain on the top after earth-movements, save in very exceptional cases which need not be regarded here. For general purposes we may look upon any bed or set of beds resting on any other in our own country as being the newer bed or set.

The movements which affect beds may occur at different times. A set of beds may be laid down flat, then thrown into folds by movement, the tops of the beds worn off, and another series of beds laid down upon the worn surface of the older beds, the edges of which will abut against the oldest of the new set of flatly deposited beds, which latter may in turn undergo disturbance and renewal of their upper portions.

Again, after the formation of the beds many changes may occur in them. They may become hardened, pebble beds being changed into conglomerates, sand into sandstones, mud and clay into mudstones and shales, soft deposits of lime into limestone, and loose volcanic ashes into exceedingly hard rocks. They may also become cracked, and the cracks are often very regular, running in two directions at right angles one to the other. Such cracks are known as *joints*, and the joints are very important in affecting the physical geography of a district. Then, as the result of great pressure applied sideways, the rocks may be so changed that they can be split into thin slabs, which usually, though not necessarily, split along planes standing at high angles to the horizontal. Rocks affected in this way are known as *slates*.

If we could flatten out all the beds of England and Wales, and arrange them one over the other and bore a shaft through them, we should see them on the sides of the shaft, the newest appearing at the top and the oldest at the bottom, as in the annexed table. Such a shaft would have a depth of between 10,000 and 20,000 feet. The strata beds are divided into three great groups called Primary or Palaeozoic, Secondary or Mesozoic, and Tertiary or Cainozoic, and the lowest of the Primary rocks are the oldest rocks of Britain, and form as it were the foundation stones on which the other rocks rest. These are termed the Pre-Cambrian rocks. The three great groups are divided into minor divisions known as Systems. The names of these Systems are arranged in order in the table, and the general characters of each System are also stated.

With these introductory remarks we may now proceed to a brief account of the geology of the county.

Merionethshire in its geological formation belongs to the oldest series of rocks classified in our table. In it there are examples of the Pre-Cambrian, or as they are

	Names of Systems	Subdivisions	Characters of Rocks
TERTIARY	Recent Pleistocene	Metal Age Deposits Neolithic ,, Palaeolithic ,, Glacial ,,	Superficial Deposits
	Pliocene	Cromer Series Weybourne Crag Chillesford and Norwich Crags Red and Walton Crags Coralline Crag	Sands chiefly
	Miocene	Absent from Britain	
	Eocene	Fluviomarine Beds of Hampshire Bagshot Beds London Clay Oldhaven Beds, Woolwich and Reading Thanet Sands [Groups	Clays and Sands chiefly
SECONDARY	Cretaceous	Chalk Upper Greensand and Gault Lower Greensand Weald Clay Hastings Sands	Chalk at top Sandstones, Mud and Clays below
	Jurassic	Purbeck Beds Portland Beds Kimmeridge Clay Corallian Beds Oxford Clay and Kellaways Rock Cornbrash Forest Marble Great Oolite with Stonesfield Slate Inferior Oolite Lias—Upper, Middle, and Lower	Shales, Sandstones and Oolitic Limestones
	Triassic	Rhaetic Keuper Marls Keuper Sandstone Upper Bunter Sandstone Bunter Pebble Beds Lower Bunter Sandstone	Red Sandstones and Marls, Gypsum and Salt
PRIMARY	Permian	Magnesian Limestone and Sandstone Marl Slate Lower Permian Sandstone	Red Sandstones and Magnesian Limestone
	Carboniferous	Coal Measures Millstone Grit Mountain Limestone Basal Carboniferous Rocks	Sandstones, Shales and Coals at top Sandstones in middle Limestone and Shales below
	Devonian	Upper Mid } Devonian and Old Red Sand- Lower } stone	Red Sandstones, Shales, Slates and Lime- stones
	Silurian	Ludlow Beds Wenlock Beds Llandovery Beds	Sandstones, Shales and Thin Limestones
	Ordovician	Caradoc Beds Llandeilo Beds Arenig Beds	Shales, Slates, Sandstones and Thin Limestones
	Cambrian	Tremadoc Slates Lingula Flags Menevian Beds Harlech Grits and Llanberis Slates	Slates and Sandstones
	Pre-Cambrian	No definite classification yet made	Sandstones, Slates and Volcanic Rocks

sometimes called, the Archaean, which in the main consist of igneous or eruptive rocks. These are found stretching intermittently from Cader Idris to the two Arans, then, with a break in the Mawddach and Wnion valleys, we find them in the mountains of Ardudwy, from which they continue northwards to Festiniog. They take in by the way the Rhobell Fawr and the Arenigs.

The Rhobell Fawr is the most striking example of igneous rock to be seen. It has been formed by volcanic action upon some great primeval sea bottom long long ages ago, measured by tens of thousands of years. In its outline it shows the most extensive mass of ancient lava in the Principality, and this is surrounded for miles by hundreds of smaller examples which were once united, but have been cut off by denudation and other agencies. Cader Idris belongs to a similar series with this difference, that the summit is formed by a pinnacle of what geologists call trap rock, beneath and around which the vertical cliffs and precipices of ash range themselves.

It must not, however, be supposed that these mountains themselves are in any sense extinct volcanoes. The deep *cwms* overhung by the precipices with their streams of loose stone, are no spent craters. They are the result of great volcanic action which took place at the bottom of the sea when the whole country was under some great ocean in remote ages. Geologists assume that when this enormous mass of complicated rock made its first appearance above the water in which it had been formed, it must have presented to the eye a fairly uniform and level tableland. But after its emergence from its aqueous

Cader Idris : the Precipice

birthplace, the streams of water, rain, frost and other atmospheric agents acted upon it in course of time, and began to carve the country into the shape and form it now presents.

The softer rocks would of course become worn down much quicker than the harder, and the latter would in consequence ultimately become the superior heights of the county. In places where the softer sedimentary stratified beds enter into the composition of the surface, the hills are on the whole smooth and more uniform in shape. But where the country shows a great predominance of igneous rocks, the hills are loftier, more serrated, and bolder in their outline.

From near Barmouth through the Vale of Ardudwy to the basin of the lower Dwyryd, and also in the contrary direction from the Mawddach estuary to that of the Dysynni, we find the rocks belong to the Cambrian System. The southern part of the county from the Dovey to near the Dysynni, and again along the western slopes of the Berwyns as far as Bala, and also the north of the county, belongs to the Ordovician System. The eastern part of the county is mainly composed of rocks belonging to the Silurian System.

The Cambrian formation of rocks as seen in Dyffryn Ardudwy consists of the following strata or beds, (1) Harlech Grits, (2) Menevian beds, (3) Lingula Flags, and (4) Tremadoc Slates.

The Harlech Dome, as it is called by geologists, is a large, irregularly-oval tract lying between Barmouth and Harlech, or it may be more correctly said between

Dolgelly and Harlech and ranging northward to Maentwrog. It is occupied by unfossiliferous grits, and purple and green slates. It holds a very important place in the physical features of North Wales, being the site of the great Merionethshire anticlinal, in which the rocks dip in opposite directions like the roof of a house. On the flanks of these sloping rocks the fossiliferous flags and grits of the Lower Cambrian series are observed to rest.

The Lingula Flags are divided into three groups, called respectively the Maentwrog, the Festiniog, and the Dolgelly groups. The first is distinguished by its jointed dark-blue ferruginous slates; the second by hard micaceous flags; and the third, or the Dolgelly group, by the soft black slate which shows a black streak when scratched.

The range between the rivers Eden and the Mawddach in the neighbourhood of Dolmelynllyn has always been famous for its fossils.

Next above the Cambrian we have the Ordovician formation, which consists of the following strata—the Arenig beds, the Llandilo beds, and the Bala or Caradoc beds. This last series occupies the largest area in North Wales. It spreads by numerous undulations around the towns of Bala and Corwen; it forms the main construction of the Arenigs, the Arans, the Gader group, and the Berwyns; and in it we have the Festiniog, Corris, Abergynolwyn, and Aberllefenni slate-quarries.

The Silurian formation is seen to stretch from near Bala to the eastern bounds of the county. The special feature and interest of this formation to the native of

Merionethshire is that the term "Bala beds" is given to certain rocks found all over Wales, because they are best developed in the strata from Dinas Mawddwy by Bala to Bettws-y-Coed. The Berwyn mountains as far as the borders of Shropshire contain a special limestone known as "Bala Limestone." It is, however, not of good quality, and is not employed for building purposes, though very useful for road-making.

It is a well-known and interesting fact that the largest "fault" in the British Isles cuts through the middle of Bala Lake from south-west to north-east. This large fault, which rent asunder the rocks, occurred far back in geological time, disturbing the rocks along its course, and consequently facilitated the action of denuding agents on the beds of softer material.

The origin of Bala Lake is considered by geologists to be the work of glacial action. Professor Ramsay says that the greater part of the Silurian region on either side of the lake and of the Dee stood high above the level of the sea from remote geological times, and probably formed a wide tableland extending far to the south, and also to the east and north-east. On its edges rose the more mountainous expanse formed by volcanic rocks, splendid relics of which still remain in the peaks of Cader Idris, the Arans, and the Arenigs.

When the Dee began to flow in its earliest channel, it is clear that its present source, Bala Lake, had no existence. The river at that time must have flowed over a surface of land not less high than that on either side of the present valley near Corwen and Llangollen. The surface of Bala Lake is only 690 feet above the sea-level, while the neighbouring watershed between the lake and Dolgelly is only 200 feet higher. As the river could not flow uphill, it is clear that in that early stage of its history the valley of the Dee about Bala must have been at least 1300 to 1400 feet higher than it is now, and consisted of a mass of Silurian rocks, a great part of which has been since removed by denudation.

8. Natural History.

It is a recognised fact in Natural History that the ancestors of most of our flora and fauna arrived in the British Isles when our country formed part of the mainland of the continent of Europe, and when there was no intervening sea to prevent easy communication. Our mountains and hills abound in proofs both of a former continental and a subsequent glacial age. Before all the various species of European animals had arrived here communication with the continent became cut off. The land of the north-western districts of Europe became isolated by the submergence of low-lying plains, and the North Sea, the English Channel, and the Irish Sea were formed, causing the influx of animal life to be stopped. This is the reason, we are told, why there are more than twice as many kinds of land animals in Germany as there are in England and Wales, and nearly twice as many in England as there are in Ireland.

Some of the animals which came from the continent into Britain in the distant past have ages ago died out, either because the climate changed and so cut off their food supply, or because they were destroyed by the hunters of the Stone Age or later times. From the finds made in the Caegwyn caves near Tremeirchion, we learn that the mammoth and woolly-haired rhinoceros lived here, and from the bones found in other parts it is evident that the cave lion, cave bear, bison, reindeer, hyaena, and Irish elk were common animals. The old Welsh Triads tell us that the first settlers of Britain found it full of *Eirth a Bleiddiau, Efeinc ac ychain bannog,* "bears and wolves, beavers (or crocodiles—the word has been variously translated) and horned oxen." This may perhaps appear an exaggeration, but the truth revealed by the cave finds is beyond controversy.

Although there are many more species of beasts and birds on the continent of Europe than we have in Britain, yet both birds and beasts are numerically much more common here. One reason for this is that we do not shoot or trap for food small birds of every description, as is the custom in many European countries. Game-preserving also, although it has lessened or extirpated the larger birds of prey, such as kites and buzzards, and keeps down other species such as jays, magpies, and carrion crows, provides protection for great numbers of the smaller birds, which are safe from harm during the breeding season.

When the glacial condition of our land passed away, wherever plants could grow we may be sure they were only those which could endure the cold. But when warmer climatal changes gradually ensued, such as to bring about the present state of things, the Arctic flora left the lower grounds but retained their hold on the cold flanks and summits of the higher mountains.

The mountains of Merioneth in general are over 2000 feet in height and consequently afford suitable habitats for a rich Alpine flora. The vertical precipices of volcanic ash on Cader Idris, and the greenstone on Rhobell Fawr, Rhinog Fawr, and Moelwyns are particularly rich in interesting flora, among which may be found various Saxifrages, rose root (*Sedum Rhodiola*), mountain sorrel (*Oxyria reniformis*), hairy rock-cress (*Arabis hirsuta*), hairy genista (*G. pilosa*), Welsh poppy (*Meconopsis cambrica*), northern galium (*G. boreale*), bird cherry (*Prunus Padus*), bald-money (*Meum athamanticum*) and melancholy thistle (*Carduus heterophyllus*), whilst the berry-bearing plants and heaths are very plentiful, comprising the crowberry, cowberry, cranberry, and cloudberry (*Rubus Chamæmorus*), the last being especially abundant on the Berwyns at Cader Fronwen and called by the English "knotberry," but by the Welsh "mwyar y Berwyn."

The numerous deep, shady, and secluded glens of the county are very rich in scarce plants, and not less so in mosses and other cryptogams. The littoral also, with its salt marshes and sand dunes, produces a number of interesting species. In the salt marshes may be found the marsh samphire or glass-wort (*Salicornia*), the tassel-grass (*Ruppia*), sea-blite (*Suæda maritima*), sea-milkwort (*Glaux maritima*), marsh and sea arrow-grass (*Triglochin*), sea convolvulus (*C. Soldanella*), now becoming difficult to find where once it was plentiful, and the very pretty purple sea lavender (*Statice Limonium*), also now only to be found sparingly. Among the sand-dunes occur some rare rushes, such as *Juncus maritimus* and *Juncus acutus*; the sea spurge (*Euphorbia Paralias*) is plentiful, and on the level grassy spots may be seen the exquisite ladies' tresses (*Spiranthes autumnalis*), maiden pink (*Dianthus deltoides*), common Teesdalia, fleabane, whorled Solomon's seal (*Polygonatum verticillatum*), annual knawel (*Scleranthus*), and the evening primrose (*Oenothera biennis*).

In the bogs and wet places we have the marsh St John's-wort (*Hypericum elodes*), the sweet-scented white orchis and frog orchis (*Habenaria albida* and *H. viridis*), the ivy-leaved campanula, marsh Andromeda, wild balsam (*Impatiens Noli-me-tangere*), floating water-plantain (*Alisma Plantago*), both long and round-leaved sundews, hemp agrimony (*Eupatoria Cannabina*), yellow flag iris, gipsy-wort (*Lycopus Europæus*), bog myrtle, celery-leaved crowfoot (*Ranunculus Sceleratus*), and the great spearwort (*R. Lingua*), an uncommon plant. The marshy lands in July are white with the cotton-grass, two kinds especially—*Eriophorum vaginatum* and *latifolium*. The willow-leaved spiraea forms hedges in wet places and the common spiraea known as meadow-sweet grows abundantly in the damp meadows. The railway embankments show a number of plants that are not natives, among which are red campions and scarlet poppies, whilst the American cress (*Barbarea praecox*), probably a garden escape, is found wild on the roadsides in some places together with the stately tree-mallow (*Lavatera arborea*).

Ferns are characteristic of Merionethshire. Sheltered crevices of the rocks, shady dells, extensive moors, and wild uplands have all their typical species and in many instances of a rare kind. The parsley fern (*Cryptogramme crispa*), a lover of rocky habitats, is to be found in many places. The beech fern (*Polypodium Phegopteris*) is to be found in secluded hollows, and the oak fern (*P. Dryopteris*) is plentiful in several parts. Magnificent specimens of the broad fern grow in boggy places. In old walls may be seen the hay-scented fern (*Lastraea Foenisecii*) growing in dense clusters, and in inaccessible faces of rocks *Asplenium lanceolatum* flourishes. Others observable are the little filmy ferns *Hymenophyllum*, *Tunbridgense* and *Wilsoni*, which are the tiniest and most delicate-looking of all British ferns, the green spleenwort (*Asplenium viride*), and the exceedingly rare forked-spleenwort (*A. septentrionale*) so much sought after by old-time herbalists. The *Osmunda*, which used to crown the banks of many a little stream, has become exceedingly rare through the rapacity of fern-sellers and tourists. There are also the sea fern, adder's tongue, holly fern (*Aspidium Lonchitis*), and prickly shield fern (*A. aculeatum*).

Turning to the wild animal life of the county we know from the names of some of its river-fords that the beaver in ancient times had its favourite haunts in some of our valleys, and there still survive the place names of the retreats of the bear and the wolf. The otter is plentiful, and so are the fox, badger, stoat, weasel, squirrel, hedgehog, rabbit, and hare. The polecat is sometimes seen, but the marten is now extinct, the last two being killed twenty years ago, the one at Dolgelly and the other at Corwen.

Merionethshire being a county of high mountains, steep crags, extensive moors, and deep secluded valleys, its list of birds is both varied and interesting. The salient feature, however, is that not many species found here can be accounted as native in the fullest sense, though many which have regularly made the county their home during the breeding season may be classed as residents. The summer and winter migrants are very numerous.

The raven is common ; four or five of them may frequently be seen hunting together over the grouse moors in search of sickly and wounded birds. The merlin has its favourite haunts on the Berwyns, and the peregrine falcon breeds on the rugged and sheer cliffs of the Arans. Sparrow-hawks and kestrels are common. The tawny and barn owls are to be found commonly, but the long-eared owl has become scarce. The wood-pigeon is abundant and does much mischief to root-crops and standing corn. The sheldrake is a resident of the lower Mawddach, and here too, when the tide is out, may be seen cormorants, plovers, and sandpipers of various kinds, black-headed, kittiwake, and common gulls, oyster-catchers, herons, and many other estuary-loving birds. The curlew and golden plover leave the seaside in March to breed on the moors, where the red grouse makes its home in great numbers. The finest grouse moors in Wales surround Bala Lake. The black grouse has become scarce, but it is occasionally seen on the Berwyns.

The more familiar woodland and hedge birds call for

no special remark. The nuthatch, once very rare, is becoming more common. The stonechat is seldom seen but the whinchat is a common summer visitant. The green and greater-spotted woodpeckers and the wryneck are to be heard in all the woods, and the corncrake with its harsh notes greets one in the meadows and long grass. The swampy districts during the nesting season are visited by the reed and sedge warblers, grasshopper warbler, reed bunting, Ray's wagtail, water rail, and other birds of like habitat.

Among the winter visitants is the bittern, which frequents the salt marshes of the coast; and the wild swan or hooper, which visits Bala Lake and some of the moorland pools of Llandderfel. The grey phalarope is frequently seen in the vale of Edeyrnion, and sometimes on the Upper Dee above Bala Lake. Flocks of bramblings regularly visit the county in winter; they come about the end of October and leave for their northern home in February and March. Sometimes, too, the twite or mountain linnet comes here as early as September and leaves as late as April. The green sandpiper and grey plover are regular visitors. After a heavy fall of snow it is not uncommon to see great flocks of skylarks passing over the Berwyns, to feed, most probably, on the estuary and tidal waters of the Dee. The spotted crake during its passage north and south is occasionally seen in the coast region. Coot and dabchick frequent Bala Lake in large numbers.

9. The Coast-line.

The coast-line of Merionethshire is washed throughout
its whole extent by the waters of Cardigan Bay. It
extends from the mouth of the Glaslyn on the borders
of Carnarvonshire to the mouth of the Dovey on the
borders of Montgomeryshire and Cardiganshire. At all
times the bay has the appearance of a lonely untravelled
sea, for sailors as a rule give it a wide berth. The distance, measured in a direct line as the crow flies from the
Glaslyn to the Dovey, does not exceed 30 miles, yet the
numerous windings of the foreshore, and the deep estuaries
of the Traeth Mawr, Traeth Bach, the Mawddach, the
Dysynni, and the Dovey, give the county a length of
coast-line exceeding 75 miles.

The coast presents features of a varied character, and
is interesting throughout its entire length owing to the
great contrasts of its background of mountains, which
skirt it at various distances. Some closely hem in its
foreshore, whilst others are set back by a broad belt of
low-lying stretches of sand and undulating sand-dunes.
The traditions and folklore of ancient times have much
to say of the devastation of the sea, and the submergence
of immense tracts. The proximity of the high mountains
would naturally lead a stranger to expect that the coast
would have been of a rock-bound character, but the
contrary is the case except in some rare instances.

Starting at the estuary of the Dovey we may observe
how the mountain ridges closely hem the shore at short

intervals, yet the foreshore is one broad belt of sand and sand-dunes from Aberdovey as far as the mouth of the Dysynni. The town of Aberdovey is set back on the slopes of the high ridge. Similarly the little watering-place of Towyn, a fashionable resort, stands on the rising ground at some considerable distance from the foreshore. To the north of Towyn is situated what is familiarly called

Towyn: the Dysynni

Morfa Towyn, i.e. "Towyn Marsh." Formerly this was an undrained and swampy stretch of lowland of over 300 acres, frequently overflowed by the sea. But the owners of the Ynys-y-maengwyn estates reclaimed the whole area, and though in close proximity to the sandy beach it now forms a rich alluvial tract of agricultural land.

After crossing the estuary of the Dysynni, which brings down the waters of the Talyllyn lake through a valley rich in historical lore, we pass to a shore of a pebbly character. At Llangelynin on the elevated land we have a long stretch of apparently flat and level ground. This is one of the few districts in the county which is not of a hilly nature. It is called Gwastad Meirionydd —"The Plain of Merioneth." Upon nearing Llwyngwril, a charming little village with an ancient camp close by, we find the cliffs and boulders approximating the shore. As the Mawddach estuary is approached we pass a marshy level with sedges, rushes, and rank grass, close to the shore, and only separated from it by the constructed ridge over which the Cambrian railway runs. This is known as Morfa Arthog. Surrounding the Morfa the mountains come to an abrupt termination, their high cliffs and receding spurs directing the attention to Cader Idris in the background. The landscape here is peculiarly attractive, mountain and sea coming into such close proximity. The Fairbourne foreshore is a charming stretch of sand and pebble, and the bold and rugged Arthog cliffs are beautifully clothed with foliage and trees.

The estuary of the Mawddach is crossed by a remarkably fine trestle bridge of timber, half a mile in length, over which the railway passes. This also forms a gangway for pedestrians, and is in reality the promenade pier of Barmouth, stretching for half a mile over the "aber." The view inland is one of the finest in Wales, with Cader Idris to the right, Diphwys to the left, the estuary below, and the Arans beyond. At the Barmouth end, where the

Barmouth Estuary

course of the river lies, the bridge is of iron, and is lifted by machinery when a vessel passes under, but the rest of the structure is of timber and must have denuded a pretty extensive woodland to make, for more of it is buried than exposed. Commanding the entrance to the estuary we have the small island known as Ynys Brawd—"The Island of the Brother"—traditionally the abode of a hermit in ancient days. This natural breakwater, strengthened by artificial work, protects the harbour, in which small boats find a shelter.

The foreshore at Barmouth affords a good stretch of sandy beach and the town rises in terrace fashion up the steep slope of the ridge commanding the northern bank of the estuary. Leaving Barmouth and proceeding northwards by Llanaber, Egryn Abbey, and Llanddwywe into the Dyffryn district, we pass a wet and marshy coast which gives place to accumulations of blown sand rising into immense dunes, upon which rank grass finds luxuriant growth. This part of the coast is known as Morfa Dyffryn.

Opposite Llanbedr, at the mouth of the River Artro, lies the island of Mochras, famed for its shells, from which the Sarn Badrig, "St Patrick's Causeway"—a reef of submerged rock—stretches into the sea for a long distance. We pass northwards by Harlech, and observe the magnificent Edwardian castle towering defiantly on the summit of the steep rock a good half mile from the seashore. In former times, not far remote, the sea undoubtedly washed the base of the castle-rock. Now it is separated from it by a wide belt of level fields, banked at the foreshore by an immense barrier of sand-hills which remind us of the coasts of Holland. In these sand-hills, rank grass, rushes, and small bushy shrubs grow in abundance, so that the otherwise shifting sands are kept in bounds. Proceeding round the bend of Morfa Harlech with its famous golf links and racecourse, and passing the promontory known as Harlech Point, we enter the treacherous inlet of the Traeth Bach, the outlet of the Dwyryd river.

Ynys Giftan

Here the ever-shifting sand makes navigation to Penrhyndeudraeth most dangerous, and only very small craft attempt the channel. Much of the land on the southern side of Traeth Bach, as at Harlech, has been reclaimed from the ravages of the sea.

Skirting the northern coast of the Traeth Bach we come to the demesne lands of Castell Deudraeth, a comparatively modern mansion, and by Minffordd pass the little island called Ynys Giftan. Having rounded Penrhyn point we are in the Traeth Mawr, into which the Glaslyn flows. The sea does not now overflow the wide expanse of ten thousand acres which lies in the valley of the Glaslyn as it did a hundred years ago. This was all reclaimed in the early part of the last century, at a cost of over a hundred thousand pounds, by Mr Madocks, who barred out the sea by a huge dyke. In the making of the dyke he had as his friend and companion the poet Shelley, who came to live in the neighbourhood.

The popular traditions of the Merioneth coast tell us of an extensive tract of rich country having been submerged one stormy night by the sea. This territory is known in old Welsh records as Cantrev-y-Gwaelod— "the Lowland Hundred." There are various versions of the calamitous flood, the one found in a poem of the "Black Book of Carmarthen" being perhaps the earliest extant. The "Welsh Triads" of the Myfyrian Archaeology also refer to it as one of the three chief disasters of Britain. The version which has made it popular to English readers is that of Thomas Love Peacock, in his story, *The Misfortunes of Elphin.*

Gwyddno was King of Ceredigion (Cardigan), and his most fruitful and valuable possession is said to have been the Lowland Hundred, which was protected by an immense sea wall. Sarn Badrig, before referred to, was part of it ; so was Sarn-y-Bwch—"the causeway of the hart"—extending from the Dysynni in the direction of the Sarn Badrig. Remains of these are still to be seen at low water. The Hundred is said to have contained many towns, among which was Porth Gwyddno, one of the privileged ports of the Isle of Britain. The tradition is further perpetuated in the well-known old Welsh air "The Bells of Aberdovey."

10. Climate.

By the climate of a district is meant the average weather experienced by that country or district. It comprises its rainfall, temperature, hours of sunshine, and humidity of the air. These depend upon various conditions, among which the most important is geographical position; that is, in the first place, the latitude, or the distance of the country from the equator, and in the second place, its distance from the sea and its height above sea-level. We have also to consider the nature or character of its soil and vegetation.

Speaking generally, the nearer a country is to the equator the hotter will be its climate, and the nearer it is to the sea coast the more equable will it be. The highest temperature in the shade ever yet recorded was in the Sahara at a spot within the tropics, where the thermometer registered $127°$ Fahr.; and the lowest or the greatest cold ever experienced was at a place in Siberia, where the thermometer is said to have registered $90°$ Fahr. below zero.

The climate of our country as a whole is very much affected by the Gulf Stream and the winds that aid it and its "drift." The prevailing winds of our land blow from the west and south-west, and come laden with moisture. These winds meet with elevated land-tracts directly they reach the western shores, such as the moorlands of Cornwall and Devon, the Welsh mountains, including those which extend from Cader Idris to the spurs of the Snowdon group in this county, and the fells of Cumberland and Westmorland. As soon as the winds touch these barriers they part with their moisture and it descends in abundant rain. This is seen by referring to the accompanying map of the annual rainfall in England and Wales. It will be noticed that the heaviest rainfall occurs in the west, and that it decreases with remarkable regularity towards the east, until the least rainfall occurs on the eastern shores of England.

The rainfall along the line of our Merionethshire mountains is about the largest in the whole of England and Wales. Upon the higher mountain groups of the Arenigs and the contiguous Snowdon group the average rainfall is as high as 100 inches per annum. This is an enormous amount, but it has been exceeded on many occasions in individual years. Thus at Llyn-Llydaw, in the Snowdon district in 1908, no less than 237 inches were registered.

The heaviest rainfall in England and Wales in the year 1908 was at Glaslyn in this district, where as much as 176·6 inches were measured, while in the quarry district of Festiniog the rainfall was 81 inches. To take other parts of the county, the average at Brithdir, near Dolgelly, was 66 inches; at Llandderfel to the east of

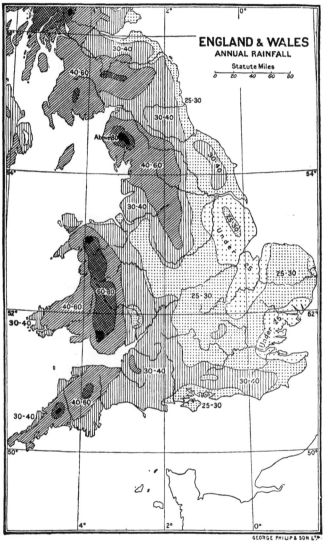

ENGLAND & WALES
ANNUAL RAINFALL
Statute Miles
0   20   40   60   80

30-40
40-60
25-30
30-40
Above 80
40-60
30-40
30-40
Under 25
25-30
25-30
60-80
40-60
30-40
Under 25
30-40
30-40
40-50
30-40
25-30

(The figures give the approximate annual rainfall in inches.)

Bala Lake it was 47 inches ; whilst at Rûg Gardens near Corwen, still further east, it was only 35 inches. In the eastern and south-eastern counties of England the rainfall did not exceed an average of more than 25 inches, whilst at Shoeburyness in Essex the lowest rainfall in the whole country was registered, being only 14·57 inches. It will thus be seen that the high ridges of Merionethshire and of the west generally may very appropriately be likened to a great umbrella, sheltering in a large measure the parts of the country to the east from the heavy and continuous rains. The months with the least rain in Merionethshire as a rule are June and July, with an average rainfall of 2·83 inches. The wettest months are November and December, with an average of 4·25 inches.

There is an important society in London called the Royal Meteorological Society which collects from all parts of the country particulars of the rainfall, the temperature of the air, the hours of sunshine, and the direction of the winds. The newspapers day by day give a summary of these particulars, so that we are able to see at a glance by means of a chart or map exactly what kind of weather has been experienced during the past 24 hours in all parts of the British Isles. At the end of the year these results are totalled and averaged, and from these we are in a position to compare and contrast the character of the climate at various places.

In 1908 the mean temperature of England was 48·9° Fahr., and of Wales 49·2° Fahr., whilst that of Merionethshire was 47·2° Fahr. Thus it will be seen that Merionethshire was below the average for both England and

Wales. The month in that year with the highest mean temperature was July with 59·1° Fahr., and the lowest was January with 38·4° Fahr.

The annual total of hours of bright sunshine in Merionethshire obviously varies from year to year. The sun is above the horizon in England and Wales for more than 4450 hours in the year, but separate districts never have the same number of hours of sunshine. No district in the whole of the country is favoured with sunshine for half the number of hours that the sun is above the horizon. The coast regions of Merionethshire as a rule enjoy more bright sunshine than do districts inland. Bright sunshine was recorded at Greenwich Observatory in 1908 for 1406 hours, and in various places in the south of England there was bright sunshine for nearly 2000 hours. But the average total number of hours of bright sunshine in England was 1498, and in Wales 1497, whilst in Merionethshire it was only 1485 hours. The sunniest months in Wales during the year are generally May and June, with an average for the last ten years of 202 hours of bright sunshine, and the month with the least sunshine generally is December, which has an average for the last ten years of 35 hours.

Sometimes it is found that places in the same county, and situated not far distant from one another, have a marked difference in climate. Configuration of the land and general aspect have much to do in producing this contrast. A hill-slope facing south receives the sun's rays more directly than does a slope facing north. Consequently we find the southern aspect is more sunny and genial than the northern aspect. It may sometimes be noticed after a heavy fall of snow in winter-time that the snow remains longer on the northern slope of the Berwyns than it does on the southern slope. Similarly in the narrow glen of the Wnion, the slopes of the left bank, which face the north, retain the snow in winter for a longer period than do the slopes of the right bank, which face the south.

The climate of Barmouth, owing to its sheltered position, is well known to be more equable and warm than that of places not far distant which are built on a contrary slope. Barmouth is completely protected from the cold north breezes and dry easterly winds by the high ridge upon the slope of which it has been built. It makes an agreeable winter resort, as do other places on this coast like Towyn and Aberdovey. One well-known author has said of Barmouth that it combines the climate of the Mediterranean with the scenery of Switzerland, which is true in a measure. The proof of this is borne out by the number of tender shrubs which flourish out of doors throughout the long winter. Among these we have hydrangeas, fuchsias, magnolias, and myrtles, which are frequently seen in the roadside cottage gardens of the peasantry outside the towns.

Humid as the climate is in general, yet the air is very healthful and invigorating. Owing to the low mean temperature snow remains on the tops of the highest mountains, especially on the Berwyns and in the sheltered crevices of the rocks, for many months at a stretch. Yet the duration of life of the inhabitants of Merionethshire compares favourably with that of the other rural counties of Wales. A very large percentage are recorded to have exceeded the ordinary span of human life, and there are a few who even reach or approximate a century of years.

11. People—Race, Dialect, and Population.

We have no written record of the history of our land before the time of the Roman invasion in B.C. 55, but we know that Man inhabited it for ages before this date. The art of writing being then unknown, the people of those days could leave us no account of their lives and occupations, and hence we term these times the Prehistoric period. But other things besides books can tell a story, and there has survived from their time a vast quantity of objects (which are daily being revealed by the plough of the farmer or the spade of the antiquary), such as the weapons and domestic implements they used, the huts and tombs and monuments they built, and the bones of the animals they lived on, which enable us to get a fairly accurate idea of the life of those days.

So infinitely remote are the times in which the earliest forerunners of our race flourished, that scientists have not ventured to date either their advent or how long each division in which they have arranged them lasted. It must therefore be understood that these divisions or Ages—of which we are now going to speak—have been adopted for convenience sake rather than with any aim at accuracy.

The periods have been named from the material of which the weapons and implements were at that time fashioned—the Palaeolithic or Old Stone Age; the Neolithic or Later Stone Age; the Bronze Age; and the Iron Age. But just as we find stone axes in use at the present day among savage tribes in remote islands, so it must be remembered the weapons of one material were often in use in the next Age, and possibly even in a later one; that the Ages, in short, overlapped.

Let us now examine these periods more closely. First, the Palaeolithic or Old Stone Age. Man was now in his most primitive condition. He probably did not till the land or cultivate any kind of plant or keep any domestic animals. He lived on wild plants and roots and such wild animals as he could kill, the reindeer being then abundant in this country. He was largely a cave-dweller and probably used skins exclusively for clothing. He erected no monuments to his dead and built no huts. He could, however, shape flint implements with very great dexterity, though he had as yet not learnt either to grind or polish them. There is still some difference of opinion among authorities, but most agree that, though this may not have been the case in other countries, there was in our own land a vast gap of time between the people of this and the succeeding period. Palaeolithic man, who inhabited either scantily or not at all the parts north of England and made his chief home in the more southern districts, disappeared altogether from the country, which was later re-peopled by Neolithic man.

Neolithic man was in every way in a much more

advanced state of civilisation than his precursor. He tilled the land, bred stock, wore garments, built huts, made rude pottery, and erected remarkable monuments. He had, nevertheless, not yet discovered the use of the metals, and his implements and weapons were still made of stone or bone, though the former were often beautifully shaped and polished.

Menhirs, Llanbedr

Between the Later Stone Age and the Bronze Age there was no gap, the one merging imperceptibly into the other. The discovery of the method of smelting the ores of copper and tin, and of mixing them, was doubtless a slow affair, and the bronze weapons must have been ages in supplanting those of stone, for lack of intercommunication at that time presented enormous difficulties to the

spread of knowledge. Bronze Age man, in addition to
fashioning beautiful weapons and implements, made good
pottery, and buried his dead in circular barrows.

In due course of time man learnt how to smelt the
ores of iron, and the Age of Bronze passed slowly into
the Iron Age, which brings us into the period of written
history, for the Romans found the inhabitants of Britain
using implements of iron.

We may now pause for a moment to consider who
these people were who inhabited our land in these far-off
ages. Of Palaeolithic man we can say nothing. His
successors, the people of the Later Stone Age, are believed
to have been largely of Iberian stock; people, that is, from
south-western Europe, who brought with them their
knowledge of such primitive arts and crafts as were then
discovered. How long they remained in undisturbed
possession of our land we do not know, but they were
later conquered or driven westward by a very different
race of Celtic origin—the Goidels or Gaels, a tall, light-
haired people, workers in bronze, whose descendants and
language are to be found to-day in many parts of
Scotland, Ireland, and the Isle of Man. Another Celtic
people poured into the country about the fourth century
B.C.—the Brythons or Britons, who in turn dispossessed
the Gael, at all events so far as England and Wales are
concerned. The Brythons were the first users of iron in
our country.

The Romans, who first reached our shores in B.C. 55,
held the land till about A.D. 410; but in spite of the
length of their domination they do not seem to have left

much mark on the people. After their departure, treading close on their heels, came the Saxons, Jutes, and Angles. But with these and with the incursions of the Danes and Irish we have left the uncertain region of the Prehistoric Age for the surer ground of History.

Although there may not be definite proof of Palaeolithic man inhabiting the remote mountainous districts of

Remains of Goidel Hut, near Harlech

this county, yet we have sufficient evidence that Neolithic man resided in many of the constituent districts which have gone to make up the county of Merioneth, for many highly polished and elaborately finished stone instruments, together with flakes and splinters of flint, have been found in and around ancient settlements, as well as in association with interments.

Their dead they buried in a crouching attitude in a cave or in a constructed stone chamber encased by an earthen mound or cairn. Probably the remains of cromlechs and stone circles, as seen in the Vale of Ardudwy, are the work of these people.

The survival of this Iberian race is to be seen in the Welsh language of our own time, in the sequence of the words in a sentence. In Welsh the place of the verb is before its subject, as *Darllennodd Owen y Llyfr*—"Owen read the book." Another survival is the relative position of the adjective and noun, the former following the noun it qualifies or limits, as *Ty mawr*—"a large house"; *Bachgen da*—"a good boy." Philologists tell us that all languages of the Aryan stock except Welsh and its cognate languages have their nouns, verbs, and adjectives in a different order, as we have them in the English language of to-day.

In the fourth century before Christ, as already stated, the second wave of the Celts, called the Brythons, made their appearance in our country and settled here. They gradually won from the Goidels the plain districts of the land, and pushed themselves into Wales by way of Shropshire and the southern parts of Cheshire and Denbighshire, so that by the time of the Roman conquest the tribe known as the Ordovices came into possession of the greater part of our present Merionethshire as far as the Dovey. It is rather difficult to define with precision the progress made in civilisation by these early peoples before the coming of the Romans. There are not even many survivals of Roman culture in Merionethshire. So we have to look to other counties for the best and most definite evidences. We learn from old writers that the Brythons lived in wattled huts daubed with mud and clay. They cultivated the soil and sowed corn. They possessed domestic animals such as the ox, goat, sheep, pig, horse, dog, and even fowls. They ground their corn in querns or small hand-mills, and made excellent pottery.

When we come to the Roman occupation we find military or hill stations in the county, on the lines of the roads leading from and to England and South Wales, which prove a complete conquest. But the mountainous character of the land was a great hindrance to Roman civilisation and culture as we see it in more favourably situated districts of the Principality. Merionethshire does not contain any remains of villas built in the rural parts away from the military stations, such as are found in other parts of the country. Perhaps the stationing of two very important legions at Chester and at Caerleon on the Usk is a strong proof of the insecure occupation of the mountains of the west. We have, notwithstanding, sufficient proof that the Romans held a firm hold of this territory for the purpose of working the mines of gold in the Mawddach valley and the copper mines of the Ardudwy country. In these workings there is no doubt that native labour was employed. The Brythons, contrary to other conquered nations, seem to have preserved their native language. In all other countries the Latin speech superseded the native, and the latter gradually died out of existence. This is observed particularly when we come down to medieval times. In the countries of the continent where the Romans held sway for a long period of years the languages spoken were Romance sprung from the Latin ; but the contrary was the case in Wales, Cornwall, and Strathclyde.

The influence of the Romans upon the Brythons is noticeable in the loan words from Latin which have entered into the composition of the Welsh speech. This is seen in such words of a military character as saeth (arrow), mur (wall), ffos (trench), castell (castle), pont (bridge), pabell (tent). In the matter of building houses the Brythons copied the Roman style, and we have such loan words as ffenestr (window), pared (partition), ystafell (chamber), colofn (pillar), and trawst (beam). Things used in the house which were new to the Brythons have provided also our Welsh vocabulary with cyllell (knife), dysgl (dish), cradell (gridiron), phiol (bowl), canwyll (candle), lleitheg (couch), and cadair (chair), as well as many others.

The influences of Saxon times were only trivial, and were not felt to any appreciable extent in this remote territory, notwithstanding the fact that the formidable Mercian kingdom was on the border. Neither did the Normans exert the same influence here as we find they did in the sister county of Montgomery. The people remained for all practical purposes a free, independent, and unmixed race until the conquest of Wales by Edward the First, when its present chief component parts were incorporated into a county or shire.

The people, Welsh in speech and sentiment, speak

the Gwyneddian dialect of the Welsh language, which differs much from the Gwentian and Demetian dialects of South Wales. It has even many points of difference from the Powysian dialect of the neighbouring county of Montgomery. The native inhabitants seem to have been less subject to influence by the English-speaking peoples of the midlands of England than those of Montgomery-shire, or even than those of parts of Denbighshire, for they retain their Welsh characteristics and love of old Celtic traditions more markedly perhaps than any county in Wales.

The population of Merionethshire at the present time is not as large as that of most other counties of North Wales; in fact, with the exception of Montgomeryshire, it is the most sparsely peopled county of the northern half of the Principality. When the last census was taken in 1911 there were 45,565 persons in the administrative county, or 69 to the square mile. This is a decrease of 3287 from the census of 1901, mainly to be accounted for in the Festiniog and Corris quarry districts by the recent depression in the slate trade. There has, however, been a great increase in the last hundred years, for in 1801 the population only reached 29,506. This increase has mainly taken place in the slate-quarrying and mining neighbourhoods, and in the watering places of the coast at Barmouth, Towyn, and Aberdovey.

The census of 1911 shows that there were more females in the county than males. The former numbered 23,763 while the latter were 21,802. These lived in 11,183 inhabited houses, of which 4051 were houses of less than five rooms. The main occupation of the people appears to be agriculture, in which 4561 males and 574 females are engaged. The men and boys engaged in the quarries number 3895, and those engaged in the copper and gold mines 479. In the building and constructive trades there were 1207. The Welsh flannel factories at present employ less than a hundred hands.

It is interesting to observe that in the census of 1901 (the last figures available on this point) the number of native-born folk of the county enumerated in other places in the British Isles exceeded the population to the extent of 5650. Of the 48,852 enumerated in the county of the ages of three years and upwards, 23,081 were monoglot Welshmen; 19,674 were bilingual, speaking both English and Welsh; while 2825 spoke English only.

12. Agriculture — Main Cultivations, Woodlands, Stock.

The mountainous character of the county interferes considerably with cultivation. The valley districts, however, are generally fertile, and praiseworthy efforts have been made to improve the quality of the soil. The slopes of the mountains are frequently boggy and very bare, and consequently provide poor pasturage for sheep and cattle. Notwithstanding such natural draw-backs, great improvements have taken place on the large estates within recent years, by a systematic scheme of drainage, and enclosing of waste lands. The quality of the soil has been much improved by regular courses of various kinds of manures, with a generous application of lime. The result of this now is that we see many of the hilly slopes covered with herds of small black cattle, while with occasional feeding on enclosed lands, the small Welsh mountain sheep, as nimble as goats, do well in the summer-time on the scanty herbage of the mountain slopes.

The returns of the Board of Agriculture record that there were in 1912 455,789 sheep in the county. This enormous number is not exceeded by any county in Wales save Breconshire and Montgomery. Merioneth-shire may thus be considered mainly a great sheep-breeding district. The quality of the mutton is admitted on all hands to be excellent, hundreds of carcases being sent weekly to the larger towns. The number of cattle fed in the county in 1912 was 36,937.

The landowners of Merionethshire, as is well known, have greatly encouraged the farmers in the improvement of the land. Much money has been spent in reclaiming the turbaries and wastes by drainage. The Ynys-y-maen-gwyn estate near Towyn has been almost entirely won from waste moors and wild and barren uplands. The embanking of the Traeth Mawr in the early years of the last century reclaimed an immense expanse on the left bank of the Glaslyn from the overflow of the sea. Similarly in the Dwyryd valley much land has been gained by embankments and careful drainage. This land, formerly a barren marsh, has been completely transformed into farms of luxuriant fertility. The embanking of the Dysynni was done at an enormous cost by the owners of the Peniarth estates. In the north-east, too, on the Rûg estates we find that much reclamation by proper drainage of the wet, peaty, and argillaceous soil has taken place. The county in general to-day presents an appearance very different from that

The Glaslyn River: Snowdon in the distance

when the Rev. Walter Davies (Gwallter Mechain) made his report upon it in the early years of the last century. He then spoke of it as possessing an immense area of irreclaimable wastes.

The acreage under crops is greatly on the increase, and by the last returns there appear to have been 151,945 acres under cultivation, which is over one-third of the whole superficial area of the county. It must not, however, be forgotten that this estimate does not include mountain and heath-land.

The number of acres devoted to the raising of various kinds of corn in 1912 was 13,650, of which 8986 acres were under oats and 4081 acres under barley. The quantity of wheat-land was very small and did not exceed 519 acres. The growing of root-crops is on the increase and more attention is being bestowed upon them; turnips, swedes, mangolds, cabbages, potatoes, and vetches accounted for 3250 acres, whilst clover, sainfoin, and grasses under rotation took up 10,526 acres. Much land within recent years has been laid down to pasture. This is due partly to the increased cost of labour and partly to the diminished value of corn. The mountain and heath-lands devoted to grazing is about 196,000 acres.

Of the cultivated land it is of very great interest to read that the land occupied by tenant-farmers is close upon 142,000 acres, whilst that in the immediate hands of owners is scarcely 11,000 acres.

Woodlands are of three kinds. First come the coppices or woods which are cut periodically, and of these there are 468 acres; then there are plantations, that is woods planted within recent years, which comprise 314 acres; and lastly we have the forests or woods of long standing which cover 15,912 acres. There has been an increase during the last ten years of one thousand acres of land under wood. The alders are periodically cut down for clog-making.

None of the farms are large, the hill-farms especially being very small. Young stock is much raised on the hill-farms, and considerable quantities of butter find their way to the Bala, Corwen, and Dolgelly markets. The farmers are a hardy race and are greatly attached to their native hills and farmsteads. They live frugally on their sixty or seventy acres, and employ but little labour outside their own immediate families, and that little only inter-mittently. They are not slaves to the plough and the harrow, as they are not much given to growing corn. Their time is largely spent in looking after their flocks of sheep and their few cattle. Usually, however, they have some hired labour at hay-time, keeping it perhaps till after the harvest.

Like the landowners who boast of a lineage going back to the time of the Welsh princes, the farmers too, in very many instances, trace their descent from the retainers of many of these old chieftains. In the report of the Welsh Land Commission we have evidences of farmers whose families have lived in the same valley, on the same homestead, for two, three, and even five centuries. These are not isolated instances but they are to be found in every part of the county. So great is the attachment to the old homestead that, in some cases, it was reported to be difficult to prevail upon a tenant to leave a tumble-down tenement for a new one only a few yards away.

13. Industries and Manufactures.

Next to Agriculture, the chief industry of the county is slate-quarrying, which has several thousands of men and boys engaged in it. The chief centre is Festiniog, where over 3000 hands are employed. Corris at one time had over 500 hands at work, but owing to the great depression in the slate-trade of the last few years this number has been very much reduced. At Aberllefeni about 150 hands are employed, and at Abergynolwyn some 200. The Pennal quarries are much smaller, with only 80 hands, while at Arthog there are not more than 50.

Slate-quarrying is a very old industry. It appears to have given employment to a few men in some parts of the county as early as the reign of Queen Elizabeth. In the latter half of the sixteenth century a small quarry appears to have been opened at Aberllefeni. The slates from it were used to roof the old manor-house of the place, a large half-timbered structure, of a character with the large houses of the Tudor period. To what extent a systematic working of the quarry was carried on it is difficult to say.

The slates of that early period were not as thin and as carefully split as they are to-day, but thick and heavy, and more like the stone tiles seen on many a cottage and farm-house in rural Wales to-day. They were of course obtained where the slate-rock cropped out at the surface, and undoubtedly lent themselves to be split easily into thick, and perhaps rough, blocks. No attempt seems to have been made to work down and scoop into the heart of the mountain where the best slate is to be found, as we now see the quarrymen do at Festiniog. On many of the hills of this county there are traces of old workings which serve to remind us of the primitive methods in vogue in olden times. From these old workings we are able to trace the old paths along which the men of those days conveyed the slate either upon their shoulders or upon small ponies to the valleys below.

Slate-quarrying in a scientific manner was first commenced in this county in the latter half of the eighteenth century. The first quarry at Festiniog was the Diphwys,

Oakeley Quarries, Blaenau Festiniog

opened in 1765. In 1800 it came to be called by a double name, by adding the name of the leaseholder of the quarry-rights, and became the Diphwys-Casson. The slates at that time were carried on pony-back in panniers to Congl-y-wal, and thence conveyed in carts to Maentwrog, where they were placed in barges and taken to Traeth Bach to be shipped.

The Bowydd Quarry was opened in 1801. It was taken on by Mr Percival in 1846, when it was known as Bowydd-Percival's. The Rhiwbach Quarry was opened in 1812. These were the earliest. Many others have since been opened, amongst which we have the Llechwedd, the Moelwyn, the Graigddu, and the Cynicht, so that the output is very great. The quarries are both open

Splitting and dressing Slates, Blaenau Festiniog

and underground, but most of the Festiniog quarries at the present time are of the latter description.

Lord Palmerston, the great English statesman, took an active interest in the quarries started by the Welsh Slate company at Rhiwbryfdir in 1816. These bear the name of Palmerston's to this day.

The main output of the Festiniog quarries is conveyed

to Portmadoc by the little narrow-gauge railway for shipment coastwise and to foreign countries. The first tramway was laid down to Portmadoc in 1833, and for thirty years horses were employed to draw the empty trucks the up journey, but on the downward journey the loaded trucks proceeded to the harbour by their own gravitation. "So it went on," says a writer, "horses pulling empty waggons and slate-trucks up, and the waggons returning the compliment by carrying the horses down again."

Steam locomotion was brought into practical use on this steep little narrow-gauge tramroad in the year 1873. The curious little double-bogie engines upon their first introduction created quite a sensation. Their ingenious adaptability for mountain climbing attracted European attention. Royal Commissions were sent from continental countries to make enquiries, under the guidance of our own Foreign Office. They came, inspected, and reported to their respective governments upon the wonderful new means of transit.

The railway is a marvellously skilful piece of engineering work, being only $23\frac{1}{4}$ inches gauge, with an average gradient of one in ninety-two. It is $13\frac{1}{4}$ miles long, and runs up-hill from Portmadoc to Diphwys, reaching a height of 700 feet above sea-level. It is computed that the cost of its construction amounted at least to £6000 per mile.

The gold-mining industry employs a considerable number of hands in the Mawddach valley. These mines have been worked at intermittent periods from very early

times with varying degrees of success.   The Romans in
the first centuries of the Christian era knew of the presence
of gold in the quartz rock of the Ganllwyd.   There has
been discovered on the farm of Dol-y-Clochydd on the
banks of the Upper Mawddach some of the flux of the
smelting-furnaces then in use.   Mixed with the flux were

Rhaiadr Mawddach and Gold Mine, Dolgelly

several pieces of broken pottery of Roman make.   A little
further up the stream in a small cairn there was found a
few years ago a complete and perfect earthenware vessel
of Roman make, bearing unmistakable traces that it had
been used in a smelting furnace of some kind.

Coming down to later times we have the Clogau and
the Gwynfynydd mines.   In the year 1860 a company

was promoted, which had for its chairman John Bright, to work the Clogau gold mines. The yield is said to have realised a profit of over twenty thousand pounds per annum in the first years of working, but the venture ultimately collapsed. In the nineties of the last century a great impetus was given to the search by Mr Pritchard Morgan, who carried on the undertaking himself with marvellous success. A company was afterwards formed and the work has proceeded regularly with varying fortune up to the present day.

The Gwynfynydd mines were restarted upon modern lines in 1888, and it is said that the output realised an annual profit of fifty thousand pounds for many years. These were closed down for a long time, but have recently been reopened with new machinery.

There was also a gold mine at Carn Dochan in the valley of the Lliw near Llanuwchllyn. This was on the property of Sir Watkin W. Wynn, and was worked by a company of which John Bright was also chairman. In 1869 and thereabout it continued working when other ventures in the county had ceased.

Copper mines are worked in the elevated tracts of the Llawllech ridge, in the Eden valley, and on the Glasdir near Llanfachreth, and lead mining engages many hands at intermittent periods at Towyn, Dolgelly, and Dinas Mawddwy.

Manganese ore is similarly worked at irregular intervals at Harlech, Barmouth, Moelfre near Llanbedr, Maes-y-garnedd in Cwm Nant Col, Cross Foxes, and Cwm Mynach above Bontddu.

The manufacturing capacity of the county is very small. The most important manufacture is the weaving of flannel and substantial woollen fabrics. These industries are distributed throughout the villages of the districts where homespuns are made. Dolgelly is the chief centre for this kind of woollen stuff, though Dinas Mawddwy almost rivals it in the production of flannel.

At Dolgelly, some years ago, a very superior kind of Welsh tweed cloth was manufactured, which attained considerable favour among the wholesale firms of London and Manchester. Large contracts were executed for the army, amounting it is said to £50,000 worth of fabrics annually. That epoch was the golden era of the woollen trade of the county town.

In a MS. of the late Robert Oliver Rees we learn the following interesting particulars about the trade in Welsh flannel:—"Dolgelly and its neighbourhood has been noted for several centuries for the manufacture of a kind of coarse woollen cloth called Webs or Welsh Flannel. This was formerly the principal trade and source of emolument of the town. Nearly every poor man within the town and every little farmer in the neighbourhood had his loom and made his Webs, to support himself and family. The Flannel Manufacture of Dolgelly is specifically noticed in Acts of Parliament of James I; and the Privy Council of Charles II issued two successive orders for its regulation. During the interval of peace which lasted for some years between the close of the American War and the commencement of the great European revolution of 1793, Dolgelly was

calculated to return from £50,000 to £100,000 annually
in this article only. These Webs were chiefly used for
clothing the armies. The Webs were rolled up with
machines into half-bales of about 18 yards each, two of
these bales making a whole Web. The annual sale does
not now (1848) exceed 500 bales."

The town of Bala was famous in the past for its
knitted gloves, its stockings, and its woollen caps, the
latter being known as "Welsh wigs." As early as the
opening years of the nineteenth century the Rev. Walter
Davies, in his report on the Domestic Economy of North
Wales, says that this town was the chief market for the
sale of knitted stockings and socks, as well as the centre
of a wide circuit in which they were made. The principal
hosiers of the place at that time estimated the regular
trade in these commodities at from seventeen to nineteen
thousand pounds per annum. Since that time the business
of the making and the selling of stockings in these parts
has fallen off very much.

14. Fisheries and Fishing Stations.

The sea-fisheries of the British Islands take rank
among their most important industries, and provide regular
employment for thousands of hands. It is computed
that there are regularly engaged in this industry in
our country alone nearly a hundred thousand men and
boys. The total annual value of the fish caught in the
British Islands at the present time exceeds ten million
pounds sterling.

The chief methods of fishing are those carried on by trawl-nets, drift-nets, and lines. The fish mainly caught in the trawl-nets are turbot, brill, soles, dories, and red mullet. These are called the " prime " class. In the nets there will also be found plaice, haddock, and whiting in countless numbers; these are called by fishermen, "offal," and sell at a lower price than the " prime."

Drift-net fishing is the method employed for catching mackerel, herring, and pilchards; fish that swim near the surface of the water. It is called drift-fishing from the manner in which the nets are manipulated. They are neither fixed nor towed within any precise limits, but are set out where fish are expected to be, and are allowed to drift with the tide.

The finest mackerel-fishing ground in our country extends from St David's Head in Pembrokeshire along the South Wales coast into the waters of the Bristol Channel. At times, however, great shoals of small mackerel enter Cardigan Bay in June; and are caught in large quantities in September near the mouth of the Dovey estuary. The fishermen of this little port sell much of their catches at Towyn during the height of the visiting season. The herring enters Cardigan Bay in August in large shoals, but they generally keep to the Aberystwyth portion of the coast. The Aberdovey fishermen make great hauls of herring at certain seasons.

Line-fishing is relatively insignificant when compared with the net-fishing, and yields not more than one-fortieth of the total value of fish caught on our shores.

But line-fishing is carried on pretty generally in most waters. The Welsh coast is the favourite haunt of the bass, and along the waters of the Merionethshire coast a large quantity of this fish is caught, together with plaice, turbot, and mullet.

The season for bass-fishing extends from May to September, and many are caught in the neighbourhood of Aberdovey and Towyn. There is no better water

The Beach, Llwyngwril

for bass anywhere along the coast than the tidal water of the Dysynni estuary. The fish frequently ascend the broad Mawddach estuary as far as Penmaenpool, and it is said they are sometimes taken in the upper reaches of the tidal water, especially at high tides. Sometimes they are very plentiful in the tidal limits of the Artro at Llanbedr, and during high tides they are sometimes found to travel up the creeks intersecting the Harlech

marshes. They are often taken in nets in the tidal
waters of the lower Glaslyn.

Turbot and brill are common on all parts of the
coast, especially at Barmouth. In the Dovey estuary
the turbot are caught in nets called foot-nets. A small
kind of cod is common in the Dovey estuary, and off
Barmouth. The whiting is intermittent in its visits

The Dwyryd River

between September and March, when good hauls are
taken. The flounder or fluke is plentiful in all the
estuaries, and at times is found to ascend the rivers for
long distances. Lemon-soles too are found along the
coast.

Fine lobsters and prawns are caught at Towyn. Off
Barmouth good hauls are occasionally made of grey
mullet and skate ; and the latter fish is sometimes taken

in trawls off Aberdovey. The fishermen of the Traeth Bach catch much flat-fish by spearing, and it is interesting to watch the men wade into the water up to their chests in pursuit of their quarry.

Most of the Merionethshire streams and lakes abound in fresh-water fish which afford anglers excellent sport. In the Dovey salmon, sea-trout, and sewin are plentiful, indeed for the two latter there is not a better stream in North Wales, though the Dysynni is not far behind it. In the Mawddach above Penmaenpool bridge the fly-rod is used to advantage, and good sport is experienced with the sewin. Near to Llanelltyd bridge excellent salmon, trout, and sewin fishing is obtained. The Camlan stream above Ty'n-y-Groes is well stocked with small trout and the Wnion is a good sewin stream, while its feeders are plentifully stocked with trout. It has also a variety of silvery trout with a red fin near the tail. Known to local fishermen as the "red-fin," it rarely exceeds a quarter-of-a-pound in weight, and is caught between the beginning of February and the middle of May. The Artro and Dwyryd are good sewin and trout streams. The latter in years gone by, before the slate-quarrying industry of Festiniog attained its huge proportions, was considered the best salmon stream in North Wales. The waters of the eastern watershed comprising the Dee and its tributaries also afford good fishing.

The lakes of the county are well stocked with fish of various kinds. Bala Lake has been referred to in an earlier chapter as to its shoals of *gwyniaid* and other species. The Aran is plentifully supplied with trout which rise well to the fly in May, and fish of a pound and over are often caught. The lakes at the base of Cader Idris contain plenty of trout, perch, and eels. The trout of Llyn-y-Gader are lean, unshapely specimens, and are somewhat insipid. Of all the lakes of North Wales, Talyllyn is the best natural trout water, and the fish are of a most delicate flavour. Of the lakes of the Ardudwy country, Llyn Dulyn is the best fishing water so far as numbers are concerned. The best of the Rhinog group is Llyn Perfeddan, in which the trout are beautifully golden and firm, and are of the size of a herring.

15. History of Merionethshire.

The Romans are the first people who have left us written records of our country. They succeeded in obtaining a footing in Merionethshire before the close of the first century of the Christian era. They did not find it an easy task, for they met with fiercer opposition in Wales than in any other country, and this resistance was carried on through a long series of years.

In the year 50 A.D., the two chief tribes, the Silures of South Wales and the Ordovices of our county and Montgomeryshire, acted together in resisting a general attack of the Roman legions. The Britons were led by Caractacus (more properly Caratâcus), a Brython like the Ordovices, though not of that tribe. He had fought

6—2

against the Romans for nine years, and had made his home with the Silures. As Tacitus tells us, Caractacus and his Britons were defeated at a hill-fortress in the country of the Ordovices, which was protected on one side by a river not easy to ford. Some authorities locate this spot on the Breidden Hills in Montgomeryshire, others in places not far removed from those hills, but there is no certainty to-day of the exact spot. This defeat in no sense completed the Roman Conquest. The tribes of Wales continued their resistance, and persevered in their harassing policy for nearly thirty years after this. So persistent and active were they that it is said that Ostorius Scapula, the conqueror of Caractacus, died heart-broken at the vain efforts to subdue them.

Excellent military roads were constructed here with marvellous skill, sufficiently wide in nearly all instances for wheeled chariots. Probably the Romans experienced greater difficulty in the making of roads in this county than in any other part of the country. The gradients and windings are most remarkable. These roads were made not only for purposes of administration, but also to reach the mines of lead, copper, and other minerals in the mountainous parts. There is a significant proof of this in a vicinal road which crossed from the Ganllwyd to the wilds of Ardudwy by way of the "Roman Steps," well known to tourists. These steps leading from Cwm Bychan over the pass are truly wonderfully made, forming a remarkable staircase of well-laid steps. The conveyance of mineral ore of some kind or other can be the only possible explanation of this elaborate and lengthy

The "Roman Steps," near Cwm Bychan

staircase in such an out-of-the-way place. We can almost see the long procession of British slaves toiling under these towering crags with their burdens on their way to Tomen-y-Mur, the nearest Roman station.

After the departure of the Romans in the early years of the fifth century, the chief event affecting our county was the clearing out of the Goidel or Gael by the sons of Cunedda, referred to in a previous chapter. These followers of Cunedda were Brythons like the Ordovices of Meirionydd.

During Saxon times the main districts of our county were under the rule of the princes of Gwynedd, and though some parts now in Merionethshire were in Powys, the greater influence came from the princes of Aberffraw. This was the time when all the British people came to be known as Cymry, that is, *Cym-bro*— "people of the same bro," as opposed to the *All-fro*, the foreigner.

Many princes ruled our land during the Saxon period, from Maelgwn Gwynedd to Gruffydd ap Llewelyn, the enemy of Earl Harold of Hastings fame. It comprised the age of King Arthur, around whose name there has been woven that cycle of myth and fantasy which has charmed the greatest writers of our land. Some of the Arthurian traditions still survive in this county.

In the time of Rhodri the Great, the Saxons of Mercia made many attempts to invade Gwynedd. They were every time repelled, and in 870 A.D., the Chronicles tell us that a great battle was fought near Llangollen in which the Mercians suffered terrible loss.

Under Llewelyn ap Seisyllt and his son Gruffydd
the territories of Gwynedd and Powys were ruled by
the same sovereigns, and prospered greatly. Gruffydd
extended the limits of his rule far beyond Offa's Dyke,
and his name was a terror to the Saxon. Edward the
Confessor and his powerful earls were compelled upon
several occasions to sue for terms of peace from him. So
well did Gruffydd ap Llewelyn circumvent the Danes
that they dared not attack any part of his territory. In
every sense he was the most capable ruler Wales had
yet seen, and was king over all the Welsh race. He was
murdered by the treachery of his own people in 1063,
in order to pacify the wrath of Earl Harold.

With the close of the career of Gruffydd ap Llewelyn
we enter upon the period of Norman invasion. The
Normans after their conquest of England engaged in
the task of subduing Wales and built a number of castles
with that end in view. The greatest help to them in
this was the ever-present tendency to internecine quarrels
among the Welsh. The Normans learnt to watch for
these and allied themselves with the powerful parties,
joining in every quarrel for the purpose of gaining their
own ends.

The princes who were pre-eminent in opposing the
Norman attack in the earliest stages were Bleddyn ap
Cynfyn of Powys, and Gruffydd ap Cynan of Gwynedd.
Bleddyn was killed in battle in 1075 and, upon his death,
his cousin Trahaearn ap Caradog, the ruler of Arwystli,
a part of Montgomeryshire, seized upon the sovereignty.
But Gruffydd ap Cynan, the representative of the ancient

line of Gwynedd from Rhodri the Great and Cunedda, came upon the scene.

He was assisted by the men of Gwynedd, who regarded Trahaearn, the man of Powys, as an interloper. Gruffydd's first conflict with Trahaearn was at Gwaeterw —"The Bloody Acre,"—in Glyn Cyfyng, now known as Dyffryn Glyncul in the hundred of Meirionydd. Trahaearn was defeated and was driven in headlong flight to his native Arwystli. Gruffydd pursued his advantage and forced Trahaearn to stand at Mynydd Carn on the borders of Cardiganshire in 1079, where a battle took place, in which Trahaearn was slain. This second victory decided the sovereignty of the north, and Gruffydd ap Cynan became the undisputed ruler in Gwynedd and Powys. Later, however, he was betrayed into the hands of Hugh Lupus, "the Wolf," of Chester, and Hugh, "the Red," of Shrewsbury, his bitterest enemies, and for twelve years was a captive in Chester Castle, whence he ultimately escaped to Ireland.

From Ireland Gruffydd ap Cynan returned with a fleet of twenty-three ships and landed in Anglesey. The men of Mon and Arfon immediately flocked to his standard and a little later those of Meirionydd, Ardudwy, Penllyn, and Edeyrnion. He also entered into an alliance with Cadwgan ap Bleddyn, prince of Powys, and together they harassed the Normans to such an extent that the barons ultimately appealed to William Rufus for aid.

In 1096 the Norman king came into Wales by way of Shrewsbury, and entered the highlands of Merionethshire. By the first of November he was at Mur-y-Castell,

or as it is now called Tomen-y-Mur, near Trawsfynydd.
But Gruffydd and Cadwgan from their safe mountain
fastnesses fell suddenly upon the Norman hosts and
caused William to beat a hasty retreat. He vowed, however, to return the following summer and exterminate
every Welshman in the land.

Rufus did return the following year as he had pur-
posed, but he was again compelled to quit Wales defeated,
having suffered great losses in men, horses, and treasure,
and having slain scarcely a man of the nation he had
vowed to exterminate.

After this there was peace in the land for a time, and
the English were content to leave Gwynedd and Powys
to be ruled by their native princes. Friendly intercourse
in time took place between the two nations, and Cadwgan
ap Bleddyn ultimately married the daughter of a Shropshire
baron.

Gruffydd ap Cynan was succeeded by his son Owain
Gwynedd, a prince of great prowess. He reigned at a
time when Wales presented a united front to English
encroachment and most certainly commanded due respect
from the kings of England. The age of Owain Gwynedd
was an age of great princes, and his influence lived through
the rule of his grandson Llewelyn ap Iorwerth (surnamed
the Great) and that of Llewelyn ap Gruffydd, the last
prince of Wales.

Owain Gwynedd was a great strategist. His success
in frustrating the attacks of Henry II at various times,
and especially in the year 1166 on the Berwyns near
Corwen, proves him to have been one of the most capable

and intrepid captains of his time. He died in the year 1169 after a reign of thirty-two years, having successfully defended his father's realm and anticipated that union of Wales which his grandson Llewelyn the Great firmly established.

Llewelyn ap Iorwerth, surnamed "the Great," grandson of Owain Gwynedd, fills an important place in the annals of Wales of the thirteenth century. He dispossessed Gwenwynwyn, prince of Powys, of the commote of Penllyn and its castle of Bala, of which the famous "Tomen" was probably the castle mound. Llewelyn is said to have fortified this castle to prevent inroads into Meirionydd and Ardudwy.

Llewelyn married Joan, daughter of King John of England, but he was often at variance with his father-in-law, though his wife frequently proved an able peace-maker between them. In the obtaining of the Magna Charta Llewelyn took a very active part, and the Welsh benefited equally with the English in its provisions. All the lands which had been taken from them in Wales and the Marches were restored to them. Other ordinances conferring special liberties and privileges upon the Welsh princes were obtained by Llewelyn, who carried great influence in the councils of the barons. Acting upon the restoration of the land by the Great Charter, Llewelyn called together all the Welsh princes to define the limits of each separate territory. This council met at Aberdovey in 1216, and the Welsh chieftains are said to have returned home acknowledging Llewelyn as their suzerain lord.

The last of the native princes was Llewelyn ap Gruffydd, the grandson of Llewelyn the Great. He came to the throne in 1255 and for twenty-six years proved himself a most capable and efficient ruler. He dominated the country much as his grandfather did, and succeeded in obtaining a much wider territorial influence as well as a prouder title. He played a great part in the politics of England and Wales of the thirteenth century and succeeded more than any other prince in infusing a patriotic spirit into the life of his countrymen. He met his death in a most unexpected manner, not at the head of his army in a great fight, but alone when engaged upon some private errand in a part of the country far from his native domains. With him died the spirit of Welsh independence. There was no one to wear his mantle. But the spirit of Welsh nationality was not dead. It had been made an enduring principle of Welsh aspiration by its last prince.

16. History—Later Times.

In a little more than a hundred years after the fall of the last Llewelyn occurred an extraordinary rising, led by Owain Glyndwr of Glyn Dyfrdwy, which occupies an important place in the annals of Merionethshire. Bold, clever, and versatile, Glyndwr at once captivated and held spell-bound the zeal and aspirations of his countrymen in all parts of Wales. They flocked to his standard in immense hosts wherever he chose to unfurl it. He appeared at a time in the history of the nation when the people were suffering from the harsh, unjust, and cruel laws made in the reign of Henry IV, laws intentionally passed in order to crush out of existence every semblance of Cymric aspiration and nationality.

The direct and immediate cause of the rising was due to the enmity between Glyndwr and his nearest neighbour, Reginald, Lord Grey of Ruthin, the Marcher Lord, who in the exercise of his duties as the King's representative purposely delayed serving Glyndwr with the King's commands. The inability of Glyndwr to honour the commands was looked upon as an act of treason to the Crown. Lord Grey of Ruthin now obtained what he had long desired. By permission of the King he attacked the Welshman's homestead at Sycharth and seized upon the estates of Glyn Dyfrdwy. This was done in so sudden and unexpected a manner that Glyndwr had barely time to escape. He soon retaliated, however, by burning Lord Grey's castle at Ruthin to the ground, and the affair at once became one of national importance. The people came to him in thousands to Corwen from all parts. From Edeyrnion, Penllyn, Ardudwy, and Dolgelly they flocked to his standard, and hailed him as "Owain, Prince of Wales."

Henry Hotspur, a son of the Earl of Northumberland, was at that time Justice of North Wales and Constable of its chief castles. He was commanded by the King to take action forthwith, and accordingly in May, 1401, he proceeded to Dolgelly with a strong military force. At the foot of Cader Idris he met with the forces of

Glyndwr.   A severe but undecided conflict took place, in which the followers of Glyndwr fully held their ground.   Hotspur did not attempt to renew the attack, nor did he pursue Glyndwr farther, but quitted North Wales and resigned his offices of Justice and Constable.

Glyndwr next met his enemy Lord Grey, the battle taking place on the slopes of the Berwyn facing the Vyrnwy.   The Lord of Ruthin not only lost the day, but was also made a prisoner.   After several months in the castle of Dolbadarn he was released, having consented to pay the heavy fine of ten thousand marks, an enormous sum for that time.   The King, however, became surety. He had also to take oath that he would never bear arms against Glyndwr again.   The King three times led an expedition against Glyndwr, and three times was beaten back discomfited.   Shakespeare refers to this in *King Henry IV* when he makes Glyndwr say :—

"Three times hath Henry Bolingbroke made head
  Against my power.   Thrice from the banks of Wye
  And sandy-bottomed Severn have I sent
  Him bootless home and weather-beaten back."

(Pt i, iii, i.)

Glyndwr summoned a parliament of his countrymen in 1403, which met at Machynlleth.   To this parliament the majority of the nobility and gentry came in very large numbers.   Subsequently, in 1404, another parliament met at Dolgelly, from which overtures were made to the King of France "as a brother and an equal," which were responded to in due course by the French King.

The Gateway, Harlech Castle

The castle of Harlech had been seized upon early in the rising by Glyndwr. He maintained possession of it almost to the end of his brief but brilliant career. His family lived at the castle, together with his son-in-law, Edmund Mortimer, who died during the siege of the castle by Henry IV's forces. In 1409 an attack was made upon Harlech, led by Gilbert and John Talbot for the King; the besiegers comprised one thousand well-armed soldiers and a big siege train. The besieged were in the advantageous situation of being able to receive their necessary supplies from the sea, for the waves of Cardigan Bay at that time washed the base of the rock upon which the castle stands. Greater vigilance on the part of the attacking force stopped this and the castle was surrendered in the spring of the year. Glyndwr's wife, her widowed daughter, and the other children were made captives and taken to the Tower. But Glyndwr was not in the castle when it was taken.

It is thought that he died in 1416, but no one knows with certainty when or where. He appeared like a meteor and was gone as suddenly. His greatness, however, is shown in the effort which he made to create out of the chaos of the time in which he lived a nation with high ideals. His letter to the King of France proves him to have been an ardent lover of his country, of which he laboured to secure the independence. He desired also to establish two universities in Wales, one in the north and the other in the south. All these ideals vanished with his disappearance, and have been realised only in part some centuries after his death.

The seventy years which elapsed from the disappearance of Owain Glyndwr to the coming of the Tudors were dark days in the history of Wales. When Edward IV ascended the throne after the victory of Mortimer's Cross, in which he was assisted by a fine body of Welshmen, he set about obtaining possession of all the castles. Only a very few of these strongholds held out, among which was Harlech, where young Henry Tudor, Earl of Richmond, had found a temporary home.

The King sent an army to take it commanded by that famous Welshman, Sir William ap Thomas of Raglan, otherwise known as Sir William Herbert, Earl of Pembroke of the second creation. It was reduced by famine and young Henry Tudor was made prisoner and conveyed to Sir William's own castle at Raglan. He was not a captive for long, however, as at the battle of Banbury, in 1469, Edward IV was defeated, and Sir William Herbert became a prisoner and was afterwards beheaded. Jasper Tudor obtained the release of his nephew, and for the next two years the young man probably spent much of his time between his uncle's domains and those of his friends the Vaughans of Corsygedol.

After the defeat of the Lancastrian party at Tewkesbury in 1471, Jasper Tudor and the young Earl of Richmond fled to Brittany, where they remained for fourteen years. In the manuscript history of the Vaughans of Corsygedol in the Mostyn Library it is recorded that Griffith Vaughan of Corsygedol in the reign of Edward IV, or that of Richard III, erected a house at Barmouth that he might the more easily be in communication with Jasper Tudor and his nephew. This house was called Y Tŷ gwyn yn Bermo—"The white house at Barmouth." This manuscript further states that young Henry Tudor and his uncle escaped to France from this place.

Henry Tudor landed at Milford Haven in 1485, and was welcomed by many thousands of his countrymen. They marched in great array to Market Bosworth, and there on Bosworth Field was fought the decisive battle which placed a scion of the ancient line of Cunedda upon the English throne. The men of Meirionydd and Ardudwy were there in great numbers, led by the Lord of Cwm Bychan, in whose honour the Cymric air, "Ffarwel Dai Llwyd," was composed.

17. Antiquities.

As stated on a former page no relics of Palaeolithic man have as yet been discovered in Merioneth, but when we come down to the next period, the Neolithic, we have quite a wealth of evidence of various kinds, not only in highly polished celts and other implements, but also in monuments, such as stone circles, standing stones, barrows, cromlechs, tumuli, and encampments of a very striking character. It would be difficult to name a district of Wales where more numerous or more important vestiges of this kind exist than in the commote of Ardudwy.

The most interesting and unusual specimen of stone implement found in Merionethshire is the beautiful and elaborately-finished hammer-head discovered at Maesmawr, near Corwen, more than fifty years ago, and particularly noticed by Sir John Evans in his great work on *Ancient Stone Implements*. It is covered with reticulated ornamentation worked with exact precision, and must have cost much labour, and the perforation for the handle is formed with singular symmetry.

Several flakes or chippings of flint having very fine edges were found in 1854 mixed with burnt bones and ashes in a cist of a stone circle near Llanaber. These flints were found in a district which has produced no native flint, and they were undoubtedly buried with the Neolithic warrior who owned them, in the belief that they would prove of use to him after death. In the cist of a carnedd on Ffridd Eithinog, near Corsygedol, there were also found flakes or chippings which were not of flint, but of a hard siliceous grit. These were also mixed with burnt bones and ashes, proving that the early people of this territory cremated their dead. In other parts of Merionethshire similar hard-stone flakes have been found.

The Ardudwy country is full of the larger remains of the New Stone Age, comprising chambered cists encased by immense cairns and surrounded by stone circles. We have cromlechs and chambered barrows in plenty. On the slopes of the hills running parallel to the coast, at different elevations, are innumerable remains of dwellings, enclosures, graves, and fortified strongholds of the prehistoric people. These strongholds, almost without exception, appear to be connected with the lines of communication and passes, and were undoubtedly places of safe retreat in case of danger. Of the chambered barrows and cairns many have been ruthlessly destroyed; the stones of them having been used to build the substantial boundary walls seen in the neighbourhood. But what remains of them are very characteristic as structures of the Neolithic Age.

Perhaps the most interesting of all are the Carneddau Hengwm, about a mile to the east of Egryn Abbey, near Barmouth, and overlooked by the remains of an immense camp known as Pen-y-Ddinas. These are two huge cairns, of which the larger is 150 feet long. They lie nearly north and south, parallel to, and near each other. Originally they contained several cists or chambers surmounted by capstones. Archaeologists state that nowhere throughout Wales or England do there exist any monuments of the Neolithic Age so striking as these of Carneddau Hengwm. The camp on Pen-y-Ddinas was the one which afforded shelter to the village settlement connected with these carneddau. It commands two old routes or trackways running north and south, the one between the fortress and the coast, and the other, thought to be the older route, running inland towards Corsygedol.

About two miles to the north is another old camp called Craig-y-Ddinas, 1164 feet above sea-level. The little stream Ysgethin washes the foot of the hill, and Corsygedol is just two miles away in the valley below. It is an excellent example of a stone fortress, and occupies the entire summit. Immense boulders have entered into its construction, and it has an entrance by a sloping passage with walls on either side. At the distant end from the entrance there are two circular apartments. How thickly peopled the neighbourhood was is shown not only by the remains of habitations and circles, but also by the numerous cromlechs, which, however, are in greatly diminished numbers at the present time.

The cromlechs are usually found in groups in the Ardudwy neighbourhood. Sometimes they bear extraordinary names in the vernacular. The one at Llanfair, near Harlech, is known as Coeten Arthur—"Arthur's Quoit." Although these ancient burial places are in the main the product of the Neolithic Age, it is presumed that they may have been used for interment by the people of the Bronze and perhaps the early Iron Age. In other parts of this county we find structures of a similar kind, although not so numerous as in the Ardudwy country. These are the carneddau on Foel Fynydd Isaf, near Dolgelly, and on Cadair Fronwen in the Berwyns; the latter has a huge menhir rising out of it.

In the east of the county and about a mile to the north of Corwen is the famous encampment of Caer Drewyn. It commands the vales of Glyn Dyfrdwy and Edeyrnion, and exhibits a single vallum, partly wall and partly earth, encompassed by a deep fosse. The walls appear to have been very wide and show evidences of rooms and apartments. Within the camp there are foundations of rude circular stone buildings many yards in diameter. The compass of the structure is sufficiently large to have afforded protection for a whole tribe or clan, together with their domestic animals.

To the north-east of Harlech and scattered about,

within a radius of four miles from the castle, are extensive
enclosures, more or less perfect, containing clusters of
circular dwellings.   The native folk call them Cytiau'r
Gwyddelod—"Huts of the Gael."   They are marked on

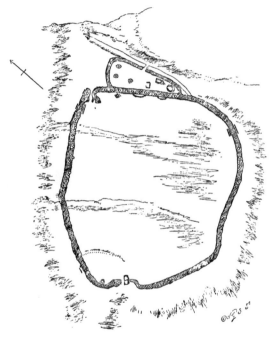

Caer Drewyn, near Corwen

the ordnance map as "Hut Circles" (see p. 61).   These
hut circles are sometimes said to be relics of the people of
the Bronze Age.   They have been much commented
upon and spoken of as vestiges of the Gael or Gwyddyl.

Archaeologists tell us that there was a primeval fortress on the rock upon which Harlech Castle stands. If such be the case it possibly formed a refuge in time of danger for the inhabitants of these circular dwellings.

The finds of bronze implements are not numerous in this county, but what have been found are very interesting. In a small mound a little to the north of the Roman fort of Tomen-y-Mur a bronze dagger-knife, $2\frac{1}{2}$ inches in length, was disinterred some years ago, together with a needle of very hard wood 6 inches in length. Both

Bronze dagger-knife found at Tomen-y-Mur

were with ashes and burnt bones in a large earthenware urn of simple design. Among the other finds have been bronze spear-heads and three dagger-blades at Cwm Moch, near Bala; a slender bronze palstave at Beddau Gwyr Ardudwy; a similar bronze palstave and dagger-blade at Trawsfynydd; a massive bronze palstave at Llanfair, near Harlech; a looped palstave and a bronze celt at Harlech. All these are in the British Museum.

Perhaps the most remarkable find of the county is the valuable gold torque discovered near Harlech in the seventeenth century, and now in the possession of the Mostyn

family. It is a wreathed rod of gold about four feet
along the curve; three spiral furrows with sharp intervening ridges run along its whole length; the ends are
plain and truncated, turning back like pot-hooks.

The antiquities of Roman times are those chiefly
associated with the military stations along the lines of
communication from South Wales and from England.

Centurial Stones from Tomen-y-Mur

(*Now in Harlech Castle*)

There are three of these stations in the county, situated
at Tomen-y-Mur near Trawsfynydd, at Caergai near the
head of Bala Lake, and at Pennal near the estuary of
the Dovey.

The station at Tomen-y-Mur was reached by several
roads, coming from various directions from the north and

from the south. It is the work of military men, as is proved by the nine centurial stones found there bearing the names of the commanders. These stones each measure 15 inches in length by 8 inches in height. Six of them are at Harlech Castle, two are at Tan-y-Bwlch Hall, and one is at Maentwrog Church. The broken pieces of Samian ware found suggest that the fort was probably built during the later years of the first century of the Christian era, possibly in the Flavian period (A.D. 70–95). The fort is situated on an eminence about half-way down the great slope which forms the eastern side of the Vale of Festiniog. The land falls from it on all sides, and a small feeder of Nant Islyn, a tributary of Nant Prysor, flows near its southern angle. It commands an extensive view of the whole country, and in shape it is rectangular with rounded corners, being 390 feet in length by 510 feet in breadth from E. to W. It encloses an area of about $4\frac{1}{2}$ acres. Two gateways, one of them with an appropriate guard-chamber, still survive. The whole is surrounded by a deep fosse and a vallum in which are the remains of Roman masonry built of hard non-local stone.

Within the compass of the fort and in the north-west quarter is a great mound of earth, from which the station takes its present name. The *tomen* is not Roman work, but is considered to be the mound of a medieval fortification reared either by Norman-English conquerors of Wales or by Welshmen copying their fashion. A little removed from the fort and near the banks of the little stream, there were unearthed the foundations of the bath-house, the invariable accompanying structure of every military fortress.

The fort at Caergai stood on a rounded hillock from which the ground falls abruptly on all sides. Its ramparts and ditch enclose a square measuring 140 yards each way and having an area of about four acres. The finds here comprise bricks, tiles, Samian ware, some blue pottery, and coins, and, in the field known as Cae Dentir, many coarse grey urns containing human bones and ashes. In 1885 there was also discovered an inscribed stone, stating that "Julius the son of Gavero, a soldier of the first cohort of Nervii, made this." It is known that this cohort was in Britain in A.D. 105, and may well have formed the garrison here.

The third Roman fort of the county is at Cefn-Caer, close to Pennal church, and near the northern bank of the Dovey. It is quadrangular in form and occupies the summit of a low hill, which rises gently in the middle of a small valley, not far from the water-side. When first discovered a pitched way led from the fort down to the river. This station is on the line of road from Llanio in Cardiganshire to Tomen-y-Mur, round Cader Idris, and was known as Sarn Elen in the vernacular. Many roads of Roman construction are known by this name in Wales. Elen or Helen was the name of a British lady of distinction, wife of the Emperor Maximus.

18.   Architecture—(a)   Ecclesiastical. Churches and Abbeys.

The county of Merioneth is not especially distinguished among the counties of Wales for its great edifices in the form of monasteries, abbeys, and cathedrals of British and medieval times. It contains no remains of the Welsh saints epoch of the sixth century, in the form of ancient côr or college, like Bangor-is-y-coed, Llantwit Major, or Tygwyn-ar-Dâf, the three great seminaries of the British period, memorable as the early home and missionary centre of primitive Christianity in these islands. Neither does it possess an ancient Gothic cathedral, like Bangor in Arfon, or St David's in the south. But many of its country churches are interesting, not so much from an architectural standpoint, as from their supposed early foundation and association with the names of the titular saints to whom they are dedicated.

The various phases of development of architecture in our country are conveniently called the Saxon, the Norman or Romanesque, the Early English, the Decorated, the Perpendicular, the Tudor, and the Renaissance styles, and a preliminary word on these various styles of English architecture is necessary before we consider the churches and other important buildings of our county.

Pre-Norman or, as it is usually, though with no great certainty, termed Saxon building in England, was the work of early craftsmen with an imperfect knowledge of stone construction, who commonly used rough rubble walls, no buttresses, small semi-circular or triangular arches, and square towers with what is termed "long-and-short work" at the quoins or corners. It survives almost solely in portions of small churches.

The Norman conquest started a widespread building of massive churches and castles in the continental style called Romanesque, which in England has got the name of "Norman." They had walls of great thickness, semi-circular vaults, round-headed doors and windows, and massive square towers.

From 1150 to 1200 the building became lighter, the arches pointed, and there was perfected the science of vaulting, by which the weight is brought upon piers and buttresses. This method of building, the "Gothic," originated from the endeavour to cover the widest and loftiest areas with the greatest economy of stone. The first English Gothic, called "Early English," from about 1180 to 1250, is characterised by slender piers (commonly of marble), lofty pointed vaults, and long, narrow, lancet-headed windows. After 1250 the windows became broader, divided up, and ornamented by patterns of tracery, while in the vault the ribs were multiplied. The greatest elegance of English Gothic was reached from 1260 to 1290, at which date English sculpture was at its highest, and art in painting, coloured glass making, and general craftmanship at its zenith.

After 1300 the structure of stone buildings began to be overlaid with ornament, the window tracery and vault ribs were of intricate patterns, the pinnacles and spires loaded with crocket and ornament. This later style is known as "Decorated," and came to an end with the Black Death, which stopped all building for a time.

With the changed conditions of life the type of building changed. With curious uniformity and quickness the style called "Perpendicular"—which is unknown abroad—developed after 1360 in all parts of England and lasted with scarcely any change up to 1520. As its name implies, it is characterised by the perpendicular arrangement of the tracery and panels on walls and in windows, and it is also distinguished by the flattened arches and the square arrangement of the mouldings over them, by the elaborate vault-traceries (especially fan-vaulting), and by the use of flat roofs and towers without spires.

The medieval styles in England ended with the dissolution of the monasteries (1530—1540), for the Reformation checked the building of churches. There succeeded the building of manor-houses, in which the style called "Tudor" arose—distinguished by flat-headed windows, level ceilings, and panelled rooms. The ornaments of classic style were introduced under the influences of Renaissance sculpture and distinguish the "Jacobean" style, so called after James I. About this time the professional architect arose. Hitherto, building had been entirely in the hands of the builder and the craftsman.

The churches of Merionethshire as a rule are small and unpretentious, but some of them contain many features which are both interesting and striking. Their dedication to the Welsh saints of the sixth century, and possibly the foundation of some of them by these ancient fathers, adds lustre to their name. Among these we have Llanelltyd dedicated to St Illtyd; Llangelynin of the supposed foundation of St Celynin; Llanfor with a double dedication to St Mor and St Deiniol; Llandrillo dedicated to St Trillo, a companion of Cadvan, an Armorican prince of the sixth century; Corwen with a double dedication to St Mael and St Sulien, two other saintly companions of Cadfan; Llandderfel, a supposed foundation of Derfel Gadarn a son of Emyr Llydaw; Llandanwg dedicated to St Tanwg; and Llanddwywe dedicated to St Ddwywe. The above churches bear the name of the saints in their designation. There are other churches which may have been founded by the early saints and are dedicated to them, but are called by other names. Among these we have Towyn founded by St Cadvan of Armorica, Llanycil founded by St Beuno, Llanaber dedicated to St Bodvan with a later dedication to the Virgin Mary, Mallwyd dedicated to St Tydecho, Trawsfynydd to St Madryn, and Llanuwchllyn to St Deiniol. The structures of this early period were undoubtedly exceedingly plain and simple. No stone edifices have survived in any part of the county. It seems probable that the first sanctuaries were made of wood and wattled plaitings, and daubed on the outside with mud and clay.

When the British prelates began to conform to Romish customs between 800 and 850, we have numerous dedications throughout Wales to Michael the Archangel, by which the churches were called Llanfihangel. In this county we have Llanfihangel-y-Pennant and Llanfihangel-y-Traethau. Afterwards come the dedications to the Virgin Mary ; among such are Talyllyn and Llanfair. Lastly followed the dedication to the apostles, chiefly the apostle Peter, and All Saints.   Such are Llanbedr and Llangar.

Llanfor Church

There is very little of Norman work in the county. The only church with any striking architecture of this style is the one at Towyn, an old cruciform structure. The nave is of a very rude description, but it is an excellent example of very early Norman work.   It is divided from each aisle by three rude semi-circular arches, on low round Norman pillars. The clerestory windows, now internally closed, are also Norman. Another church with specimens of Norman work is Llanaber, which has a chancel arch with good mouldings springing from shafts with capitals bearing foliage of Romanesque style. In

Llanaber Church

Llanegryn church there is a font of Early Norman date shaped like a cushion capital. The mode in which the upper part of the square exterior is rounded off, so as to accommodate it to the circular interior, is remarkable. The church of Llanfihangel in the adjoining parish has a similar font which, however, is of better workmanship.

It is a good Norman specimen, the bowl being square and
scalloped on its lower edge, while the stem is cylindrical
on a high square base.   At Talyllyn, too, there is an old
Norman font.

The Early English style is seen at its best in the
Llanaber church, a couple of miles to the north of

Llanegryn Church

Barmouth.   This building, perched above the sea in a
steep little graveyard, is recognised by authorities as the
most notable example of Early English in North Wales.
Though the church is small it has a nave with clerestory,
side aisles, and chancel of pure Early English work.
Within the south porch is a beautiful Early Gothic door-
way, equal to the best of that style found in Wales or

England.   The nave is divided from each aisle by five
low pointed arches springing from circular pillars, some
of which have octagonal capitals with foliage.   The
chancel has a single lancet with mouldings at the east
end, and on the north is a late square-headed window.
The clerestory of the nave is high and genuine Early
English with single lancets.   At the east end of the
aisles there are also lancets closed up.

Many of the churches of Merionethshire have some
splendid examples of the Perpendicular style.   Undoubtedly
the most striking is that to be seen in Llanegryn church,
which is situated on a slight elevation in the basin of the
Dysynni, near Peniarth.   Outwardly, it is an unpretentious structure, but it possesses a remarkable rood-loft,
considered to be the most beautiful specimen of this
kind in North Wales.   It is a work of the early
fourteenth century, and is in a perfect state, reaching
nearly to the roof.   It is finished off beautifully by two
fine vine-leaf cornices.   The great tithes of Llanegryn
parish were in olden times appropriated by the monks of
Cymmer Abbey, and they are reputed to have built the
original church, and to have brought the rood-loft here
from their Abbey.   The church of Llandderfel is superior
in its architectural character to the generality of Welsh
churches.   It is of fair Perpendicular work and perfectly
uniform, consisting of a lofty single body.   Between the
nave and the chancel is a Perpendicular rood-screen,
each compartment having foliated arches with enriched
spandrels.   The church of St Cadfan at Towyn has its
east window of three lights in the Perpendicular style.

Other churches with Perpendicular work are Llangar, Llanfor, Llandrillo, Talyllyn, Llanfair, Llanddwywe, and Llandanwg.

The ruins of monastic establishments are very few, the most famous being Cymmer Abbey. We cannot but admire the taste of the monks in choosing the most lovely spots of our land for their abode. The scenery of the

Llanfair Church

Ganllwyd, where the old Cistercians built their abbey, is most delightful in its seclusion. This religious house was founded in 1198 by Meredydd and Gruffydd, the grand-sons of Owain Gwynedd. The native folk call it by the name of Y Vaner, which some authorities interpret as *Man Ner*, "The place of God."

Its present remains prove it to have been a really fine structure, and many features still survive to tell of past grandeur.   A portion of the conventual church shows at its east end three lancet windows, and the large old refectory of the abbey, together with some parts of the abbot's lodge, form the Vaner farmhouse.

The charter of its incorporation is dated 1240.   Its

Cymmer Abbey

numerous clauses include a host of privileges, giving the monks authority over rivers, lakes, and seas, all kinds of birds, beasts wild and tame, mountains, woods, things movable and things immovable, all things upon and under the land, with full liberty of digging for hidden treasures, and unrestricted mining rights.

In a short time after its foundation we are told that a number of monks came to reside here from Cwmhir

Abbey in Radnorshire. It is, however, a mistake to say that it owes its foundation to them, as some authors state. Llewelyn ap Iorwerth, surnamed "the Great," became its patron, and confirmed to it its charter, giving certain rights to various lands. Most of the parish of Llanfachreth and the valleys to the north and west were its domains. Esgair-gawr belonged to it, as did also the parish of Llanegryn. The property of the Abbey remained in the possession of the Crown for a long time after the dissolution of the monasteries, and Queen Elizabeth gave it to her favourite, Robert, Earl of Leicester.

19. Architecture—(*b*) Military. Castles.

Merionethshire is not rich in medieval fortresses. The reason for this is that the Normans, when they occupied England and made their inroads into Wales, were not able to possess themselves of any part of this county for any considerable length of time. In some of the Welsh counties, especially in Glamorgan and Monmouth, there are more than a score of castles of Norman origin. The mountainous nature of this county made it very difficult for the Normans to build castles to overawe the Cymry, but as soon as a Norman baron settled in any part of Wales, his first thought was to build himself a fortress of stone. He never felt safe within our borders unless he had one of these strongholds to protect him from the unceasing attacks of the Welsh.

Sketch Map showing the Chief Castles of Wales and the Border Counties

When William Rufus made his two disastrous invasions of Gwynedd and was compelled to beat a hasty and ignominious retreat upon each occasion, he instructed his marcher-lords that they were to build castles in every commanding and advantageous situation. But this apparently did not affect our territory of Merioneth. The great period of castle-building in Wales came later, under King Stephen, who endeavoured to make good terms with his barons by permitting them to erect large fortresses wherever they chose.

The early Norman castle consisted of a lofty and very thick wall with towers and bastions, enclosing a wide area. Outside the walls there was generally a moat filled with water, which was crossed by a drawbridge leading to the principal gate, which was strongly defended by covering towers. Above the gate there were openings or holes through which molten lead, boiling water, or hot pitch might be poured down upon the besiegers when they succeeded in getting near. The drawbridge was raised and lowered by great chains, whilst the gate itself was a thick heavy door or a strong iron grille called a portcullis.

Within the walls was the bailey or outer court. Here were the stables, the store-houses, and part of the dwellings of the garrison or retainers. Within this again there was usually another wall protected by towers and a gate, leading to the inner bailey or court which contained the keep, a square building of great strength, in fact the citadel of the castle, the last place of retreat in case of need. It consisted of several storeys or floors, the ground floor containing no windows. It was provided with a deep well for supplying the citadel with water in time of siege.

None of the few castles of Merionethshire are of the Early Norman type. Our most famous structure, and perhaps the most famous in the whole of the land, is Harlech Castle, erected in the thirteenth century. Yet Harlech, according to the traditions of this county, goes back in its origin to much earlier times. One of its towers, known as Twr Bronwen, is a name that carries us back to the times of Bran ap Llyr. Some authorities say that the first fortress of a military character erected here was built by Maelgwn Gwynedd some time in the sixth century. In the eleventh century it seems to have been known as Caer Collwyn. Collwyn ap Tango was lord of Eifionydd, Lleyn, and Ardudwy, and lived in the time of Anarawd, King of Gwynedd, in the ninth century. Collwyn resided in a square tower of the original building, the remains of which may still be seen, for some of its walls form the base of the present structure.

The castle as it stands to-day is a fine example of the Edwardian type, erected immediately after the conquest of Wales by Edward the First. It stands on a lofty perpendicular rock, the base of which, at the time of its erection, was washed by the sea. It was utterly unassailable from the water side, whilst on the land side it was protected by a wide and deep fosse cut out of the solid rock. A remarkable feature of the castle is a covered staircase cut out of the rock, defended on the seaward side by a looped parapet, and closed above and below by

Harlech Castle

small gatehouses. This was the water-gate of the fortress, and opened upon a small quay below. The castle is a fine and commanding square structure with a circular tower at each angle. The entrance gateway is similar to that of Caerphilly Castle. Like the latter, too, it has no keep. For natural strength and grandeur of position it has no equal in Great Britain. The masonry throughout is exceedingly rough, as though hastily executed. It appears to have been completed before the year 1283, for the records of Edward the First state that Hugh de Wlonkeslow was the Constable, with an annual allowance of one hundred pounds.

Margaret of Anjou, the spirited queen of Henry the Sixth, found an asylum here after the defeat of her husband at the battle of Northampton in 1460. The south-east tower for some centuries bore her name. It was from here that she went forth with an army of Welshmen to the victory of Wakefield. When Edward the Fourth ascended the throne after the victory of Mortimer's Cross in 1461, in which he was assisted by a great host of Welshmen whose sympathies were Yorkist, only a very small number of castles continued to resist, among which we find Harlech. Dafydd ap Einion was the governor, a firm and stedfast ally of the Lancastrian party. Sir William ap Thomas, otherwise known as Sir William Herbert of Raglan, Earl of Pembroke of the second creation, was sent against it, but found himself unable to take the castle by storm. He therefore left the siege in charge of his brother Richard, to reduce it by famine. Starvation did its work, but Dafydd ap Einion would not yield except upon honourable terms, which comprised an unconditional pardon. This was granted, but Edward refused to ratify the pardon. "Then sire," said Sir Richard, "you may if you please, take *my* life in lieu of the Welsh captain's. I will most assuredly replace Dafydd in his castle, and you, Sire, may send whom you will to take him out."

Ruins of the Keep, Bere Castle

In the Civil War Harlech Castle was held for King Charles the First by Sir Hugh Pennant. It fell into the hands of General Mytton in March, 1647, and it and Raglan Castle in Monmouthshire were the last castles in the realm which held out for that ill-fated monarch.

Towering high above the Dysynni river, on a curiously detached rock, is an interesting old ruin known as Bere

Castle. It is thought to have been erected in the reign of Henry the Second. That king's discomfiture in North Wales in the year 1157 is said to have prompted the erection of this and other similar structures. It retains in its ruins some magnificent features of the Early English style of architecture, and was undoubtedly a place of enormous strength. The existing remains show it to have extended over the whole of the summit of the rock, and one apartment of it, measuring 36 feet across, is cut in the solid rock. In some parts the lines of circum-vallation consisted of loose stones piled high up on the edges of the precipice. The other sides appear to have been well defended by walls of squared stones cemented with mortar composed of calcined shells mixed with gravel.

Some authorities state that Edward the First gave the custody of Bere Castle to Robert Fitzwalter, with leave to hunt all wild animals in the country around Cardigan Bay. The castle seems before this to have been in the possession of Llewelyn ap Gruffydd, the last prince of Wales, and William de Valence, Earl of Pembroke, captured it sometime before the fall of the prince near Builth.

The Welsh Chronicles tell us that Uchtryd ap Edwyn erected a castle at "Cymmer in Meirionydd" in the year 1114. Robert Vaughan of Hengwrt, whose residence was close by, states in his writings that the castle stood on a small hillock called the Pentre, upon which in his time there was a flour-mill, a smithy, and several small tenements. We know nothing more of this old fortress;

the name only survives. Its proximity to the famous
Cymmer Abbey leads one to assume that at a later date
the monks of the old sanctuary considered its site suitable
for their abode. It may have fallen into disuse when this
took place.

Poised on a spur of the overhanging mountains in the
valley of the Lliw, a tributary of the upper Dee, is an old
ruined structure known as Castell Carn Dochan. It is
now a mere heap of stones, but we are told by reliable
authorities that it was a pre-Norman structure, a
stronghold of the princes and chieftains of Penllyn and
Ardudwy. Close by are the old workings of a gold mine,
which according to tradition was worked by the Romans
in the first century of the Christian era, when their
soldiers were encamped at Caergai in the valley below.

20. Architecture—(c) Domestic. Famous Seats, Manor Houses, Farms, Cottages.

Although Merionethshire does not possess a great
array of military structures in the form of imposing
castles and castellated mansions, yet it boasts of a wealth
of fine old manor houses and stately mansions, which
came into existence in the period immediately succeeding
the close of the Wars of the Roses, when the long epoch
of baronial supremacy came to an end. In the country
families there arose a new landed class, who were more
closely associated with the masses of the people in their

social life and aspirations, and took a greater interest in their general welfare than the old baronial families did.

Many of the older gentry of the county, of whom there are a large number, can trace their pedigree back to very remote times; and their ancient domains and homesteads have been in the possession of their families for many generations. Some of them were devoted and

The Hengwrt

careful collectors of our ancient records, and the nation owes a deep debt of gratitude to them for the collection and preservation of our earliest manuscripts, which rank amongst the most treasured records of the British Islands.

The Hengwrt is an ancient edifice nestling cosily in the trees near the ruins of Cymmer Abbey. This was the home of the Vaughans, a famous family, and the birthplace of Robert Vaughan, the well-known antiquary of the seventeenth century. His priceless collection, known as the Hengwrt MSS., is probably the most interesting and remarkable private collection in Europe. It comprises such treasures as "The Black Book of Caermarthen," "The Book of Taliesin," "The White Book of Rhydderch," the oldest version of the Mabinogion, "The Sanct Greal," and many others of inestimable value, such as the Hengwrt MS. of "The Canterbury Tales." These have been purchased for the nation by that great patriot, Sir John Williams, M.D.

Peniarth, which lies on the north bank of the river Dysynni, takes us back in portions of its architecture to the fourteenth century, but its main existing parts were built in 1700, and were added to in 1812. It was known in the earliest times as Maes Peniarth. In the reign of Henry the Fifth its owners were of the line of Ednowain ap Bradwen, Lord of Merioneth, and through his descendants in Tudor times it passed to a nephew of Lewis Owen, Baron of the Exchequer. Then it descended to a branch of the Wynnstay and Bodelwyddan families, and finally to the Wynnes, its present owners. The famous Hengwrt MSS. were safely housed here after their removal from Hengwrt, until their transference to the National Library at Aberystwyth.

Nannau, a seat of the Vaughans, stands at an elevation of 700 feet above sea-level, about two miles to the north of Dolgelly. The present house is not of later date than the closing years of the eighteenth century, but the older seat close by, now used as farm-buildings, was built in

the fourteenth century. Howel Sele, its owner in those
early days, met with a tragic end at the hands of Owain
Glyndwr. The story of the fatal fray and the finding of
Howel's skeleton in the hollow trunk of an oak tree long
years after, has been told by Sir Walter Scott in his
Marmion:

> " All nations have their omens drear,
>   Their legends wild of foe and fear,
>   To Cambria look—the peasant see,
>   Bethink him of Glendowerdy,
>   And shun the ' Spirit's Blasted Tree.' "

Corsygedol, 5½ miles north of Barmouth, the ancient
home and seat of a branch of the Vaughan family, is a
most interesting homestead of the Elizabethan age. It
has a gateway designed by Inigo Jones. One of the
family, Gruffydd Vychan, is famous for the part he took
in aiding Henry Tudor to reach the throne ; he was
subsequently squire of the body to the Tudor king. A
manuscript of Robert Vaughan of Hengwrt says that he
" was in great credit with Jasper, Earl of Pembroke, who
lay in his house at Cors y Gedol, when he fled to France
in the time of Edward the Fourth, and as some report,
Harry, the Earl of Richmond with him, who afterwards
was King of England."

Cwm Bychan is an old mansion of limited architectural pretensions, yet it is typical of the buildings so
dearly-loved by Welsh folk, and is most beautifully
situated. This was the home of the Llwyds, an ancient
family of Ardudwy, descendants in a direct line from
Cynfyn, prince of Powys, who have been the owners

without a break from Norman times. The house appears to have been originally erected sometime in the fifteenth century, and is, to-day, an old-world place rebuilt in parts about the close of the following century.

Rûg, in the vale of Edeyrnion, is a house of modern date. The old mansion, of castellated design, is now in ruins; it occupied a slight elevation not far from the present edifice. In its precincts took place the treacherous betrayal and capture of Gruffydd ap Cynan in the eleventh century. The old castle and domains were the home of Owain Brogyntyn, a prince of Powys, of the time of Llewelyn the Great.

Rhiwaedog, near the eastern end of Bala Lake, is the famous ancestral home of a long line of Welsh princes. In the twelfth century it was the abode of Rhirid Flaidd, lord of Penllyn, and at that time was called Neuaddau Gleision. The dining-room contains a magnificent cornice embodying the arms of Owain Gwynedd, whose lineal descendants were then the proprietors. In an upstairs room there are escutcheons bearing the arms of that prince, and the wolf of Rhirid. The courtyard is entered by a covered gatehouse. Above the porch of the western wing is the date 1664, but the general features of the old manor house are of much earlier date, though the eastern wing has been rebuilt within the last 150 years.

Plas Rhiwlas, the only castellated mansion in the county, has long been the home of the Price family. Its grounds have beautiful examples of various kinds of trees of great size, and the property is one of the most extensive

in the county.　Three miles from Towyn and on the
right bank of the Melindre stands Dolau Gwyn, a three-
gabled old mansion, now a farmhouse.　It is in the
Jacobean style, and its walls and ceilings are frescoed
with armorial bearings.　It was long the residence of a
junior branch of the family of Ynys-y-Maengwyn.

Plas Rhiwlas

Many other mansions of great repute are to be seen in
all parts of the county.　Some of these have as interesting
a story as those noticed above, and have their beginnings
in Tudor times or even earlier.　Among them we have
Plas Tan-y-Bwlch in the valley of the Dwyryd; Dol-y-
Moch in the same neighbourhood, with its remarkable

escutcheons of the Fifteen Royal Tribes of Gwynedd worked in the ceiling of the great hall, and its characteristic old-world chimneys; Y Dduallt, the home of the Lloyds; Tan-y-Manod; Pengwern, the county residence of Baron Newborough; Plas Dinas Mawddwy; Rhagatt,

Old Houses, Dolgelly

near Corwen, a very old place; The Hendre near Llwyngwril, now a farmhouse; Llanfendigaid near Towyn; Garthyngharad near Arthog, formerly the residence of Sir Richard Wyatt; and Hengwrt Uchaf near Drws-y-Nant, with its imposing gateway at the entrance to its park.

More or less ancient, but wearing a modern garb, are Caerynwch and Dolserau Hall near Dolgelly, the former at one time the residence of Baron Richards of the Exchequer; Glanllyn, the county seat of Sir Watkin Williams-Wynn of Wynnstay; Pale near Llandderfel, visited by Queen Victoria in the time of Sir Henry Robertson, M.P., the designer of the Dee Viaduct, and many others.

In the farmhouses a great improvement has set in during the past century. More ample accommodation is to be found in them for the family and the servants. The regular practice of whitewashing the exterior of the premises and the outbuildings is commonly resorted to in all parts of the county. This gives a clean, but artistically unattractive appearance. Farmhouses and cottages are alike built of stone and roofed with slate.

21. Communications—Past and Present.

Before we deal with the present means of internal communication in the county we may turn for a moment to consider those of the past. We have seen that the mountainous and hilly parts of the county far exceed the low-lying districts, and that there is no considerable extent of surface which may reasonably be called a plain. This has ever been the greatest obstacle to the construction of good roads. The initial expense of making them was in former times the main insurmountable barrier.

The primitive Welsh trackways and bridle-paths would pass along the declivitous glades of a valley, until at its furthest limits a mountain barrier confronted them. Our modern method would be to skirt the slopes in as gentle a gradient as possible. But our forefathers scorned such methods; they would ascend the mountain abruptly, almost in step-ladder fashion. The old roads of Scandinavia seem to have been of the same character, and they were, perhaps, as well adapted to the then state of society as ours are at present. These steep ascents covered a retreat in times of invasion or of tribal broils. Good roads would have hastened the annihilation of the liberty they so highly valued and so strenuously defended. We have some striking instances to-day of these sudden passes in Bwlch-y-Groes near Llan-y-Mawddwy, Bwlch Oerddrws near Dolgelly, and Bwlch Ardudwy above Llanbedr. Traces of many others are still with us, but they are not used as formerly, and connected with them are the ancient British trackways leading over and along the highest ridges and elevations.

The Romans were the first to construct anything in the form of a road in the strict interpretation of the term. We have referred to this in a previous chapter. There was a necessity that all parts of the country should be properly linked together by a network of roads, to enable the various divisions of the legions to pass from station to station as necessity arose. The Roman roads were excellently constructed, but after the departure of the Romans in the fifth century and the invasion of the country by various hordes from every direction, they were permitted to lapse into a state of bad repair and perhaps disuse. There was probably no attempt to maintain the roads in this county at any time during the medieval period, nor indeed until much later.

When we arrive at the Tudor period we learn that Parliamentary legislation was resorted to in order that proper means of communication might be instituted with various parts of the county and places on the border. In the reign of Queen Elizabeth we are told that the sheriffs were instructed to have bridges and causeways made between Shrewsbury and the county town. But travelling was of necessity difficult then, even in the most favoured parts of the kingdom. Macaulay tells us that at the close of the reign of Charles the Second travelling was exceedingly laborious. Goods and passengers were often carried on trains of pack-horses. Six horses were frequently needed to draw a gentleman's carriage to places only a few miles distant from London. What must the conditions have been in places so distant as this county ! He gives us an example in the journey of a viceroy going to Ireland by way of Chester and Holyhead in the year 1685. The viceroy was five hours travelling from St Asaph to Conway a distance of only fourteen miles. Between Conway and Beaumaris he was forced to walk a great part of the way, and his lady was carried in a litter. His coach was, with much difficulty, and by the help of many hands, brought after him entire. In general, carriages were taken to pieces at Conway, and borne, on the shoulders of stout Welsh peasants, to the Menai Straits. One chief cause of the badness of the roads seems to have been the defective state of the law. Every parish was bound to repair the highways which passed

through it. The peasantry were forced to give their
gratuitous labour six days in the year. If this was not
sufficient, hired labour was employed, and the expense
was met by a parochial rate.

Matters were very little better at the opening of the
eighteenth century, and for many decades after this, until

Old Coach Bridge, Dinas Mawddwy

the Turnpike Trusts were formed by Act of Parliament.
The first Act was that of 1758, which affected only a
very small part of this county. There followed another
in 1768, entitled an "Act for repairing and widening
several roads in the counties of Salop, Montgomery, and
Merioneth." This Act had reference only to the chief
roads leading from the east in continuation of the main

road from Shropshire. But nearly fifty years after this, when the Rev. Walter Davies (Gwallter Mechain) made his report to the Government, the roads were in a sorry state. He says of North Wales in general that there were comparatively but few miles of "travelable roads within the whole district. Coal for fuel, and lime for manure could not be carried in quantities to any great distance."

The modern improvement in road construction dates from the opening years of the latter half of the last century. At the present time nothing strikes the tourist with greater admiration than the excellent public roads now existing in all parts of the county. Some of these have been constructed at immense labour and expense. The circuitous road of the vale of Mawddach from Barmouth to Dolgelly through Bontddu and Llanelltyd is a magnificent piece of work. It has been engineered with great skill and in some parts of its course it has cost as much as two guineas a square yard to make.

Another road which entailed great expense, but is of great utility, is that from Pont Aberglaslyn to Penrhyn-deudraeth. Before its construction, travellers were obliged to skirt the mountain side, or what was equally incon-venient and more dangerous, to seek a guide to lead the way by the winding route over the Traeth Mawr sands. As this was passable only at low tides, it often entailed detention for a day or night. The present road, winding around an amphitheatre of mountains, is one of the most pleasing in the whole of Wales. The main road from Barmouth to join it passes through Dyffryn Ardudwy by

Llanbedr and Harlech along an elevated level, in full view of the sea.

From Dolgelly the main road to Bala and Llangollen runs almost parallel with the railway. It passes over Drwsynant and crosses the main watershed to descend to Llanuwchllyn. It then skirts the Bala Lake and passes through the town of Bala, Llandrillo, Corwen, and Glyn Dyfrdwy to Llangollen. The county town is connected with Festiniog in the north by the steep-gradient road which passes through the Ganllwyd and Trawsfynydd, to descend into Maentwrog, then ascend again by a winding gradient into the slate-quarrying area. Festiniog is also connected with the east of the county by the main road to Bala, which crosses the Migneint to enter the vale of Tryweryn and proceed by Frongoch and Rhiwlas to join the Dolgelly and Llangollen road.

Bala and Dolgelly have each direct communication with Dinas Mawddwy. The road from the former skirts the southern shores of the lake and then ascends the Berwyn mountains at Rhydybont, passing over Bwlch-y-Groes pass. The road from Dolgelly crosses the ridge after a steep ascent to Bwlch Oerddrws. At Gwanas this road branches off to Towyn and Corris, and skirts the base of Cader Idris, passing by Tal-y-llyn Lake and Abergynolwyn to the seaside resort. There is also a road from Dolgelly along the left bank of the Mawddach to Penmaenpool, then in a direct line to Arthog, where it gradually ascends to Llwyngwril, and proceeds along the coast to Towyn and Aberdovey.

The county is well served by railways both coastwise

and inland. The coastal district is served by the Cambrian Railway. The main line from Montgomeryshire in one direction, and Aberystwyth in the other, enters the county after leaving Dovey Junction, and proceeds through Aberdovey, Towyn, Llwyngwril, and Barmouth Junction, from which it crosses the Mawddach Estuary by the long trestle bridge to Barmouth. It proceeds thence along the coast through Dyffryn, Harlech, Penrhyndeudraeth, Minffordd, Portmadoc, and Criccieth, to Avonwen and Pwllheli.

At Minffordd the Festiniog narrow-gauge railway meets the Cambrian. It passes through Penrhyn, Tan-y-Bwlch, and Tan-y-Grisiau, and terminates at Festiniog. The Cambrian connection with Dolgelly leaves Barmouth Junction and passes through Arthog and Penmaenpool.

The Great Western Railway enters the county from Ruabon and Llangollen on the east at Berwyn, and thence proceeds by Glyndyfrdwy, Carrog, Corwen, Llandrillo, and Llandderfel to Bala Junction. It branches there, one connection leading through Llanuwchllyn, Drwsynant, and Bontnewydd to Dolgelly with running powers to Barmouth. The other connection proceeds to Blaenau Festiniog through Bala, Frongoch, Arenig, Cwm Prysor, Trawsfynydd, Maentwrog, and Manod. This route from Bala to Festiniog is probably the wildest and most impressive stretch of railway travelling in the whole of England and Wales.

The London and North Western Railway enters the county in the north, having come from Llandudno Junction through Bettws-y-Coed, and passes by a very

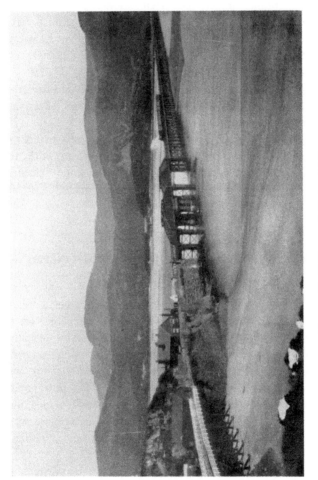

Barmouth Bridge and Cader Idris

long tunnel under the Arenig Fawr to terminate at Blaenau Festiniog. The Denbigh, Ruthin, and Corwen branch of the same railway enters the county at Derwen and passes through Gwyddelwern before arriving at Corwen.

At Towyn a small branch line proceeds to the slate-quarries of Abergynolwyn by Rhydyronen, Brynglas, and Dolgoch villages.

Another little branch line of a narrow-gauge description leaves Machynlleth and communicates with the Corris quarries, terminating at Aberllefenni, whilst a similar line leaves Cemmaes Road and proceeds up the Dovey valley to Dinas Mawddwy.

22. Administration and Divisions— Ancient and Modern.

In order to understand properly the present administration and government of the county it will be best to consider its divisions before the Tudor Period, when it was made shire ground according to English laws. Throughout the history of our particular territory we find that whenever changes were made in its government and administration, great care was exercised in adapting them to established customs, so that old institutions were not uprooted and supplanted by the new conditions introduced.

Meirionydd under the native princes constituted a third part of the territory of Gwynedd, and it was ruled by its prince from Aberffraw. We learn from Sir John Price's *Description of Wales*, of the reign of Henry VIII, prepared in the form of a petition from the Welsh people for the purpose of effecting the union of Wales with England, that Meirionydd consisted of three cantrevs or hundreds having three cwmmwds or commotes in each ; these were Cantrev Meirion, Cantrev Penllyn, and Cantrev Arwystli. The three commotes of Meirion were Talybont, Pennal, and Ystumaner ; of Penllyn, Uwch-meloch, Is-meloch, and Migneint ; of Arwystli, Uwchcoed, Iscoed, and Garthrynion.

The cantrev or hundred was the division of a country next in size to a shire, and has been generally recognised to mean, although perhaps not in a strict sense, one hundred free families. When we look at the great disparity in size of some of the hundreds, we naturally conclude that the term cantrev was not always limited to the same definite number, but meant a group, or assemblage of *trevs*.

The *trev* signified the family, not exactly in the sense that it was limited to the immediate bond of relationship of parent and child, but a clan or assemblage of blood-kindred, who associated themselves with the head of the family. The *cant-trev* or hundred was thus a joint family, and implied a common descent and brotherhood of its members. Every separate cantrev had its own hereditary head or leader. The land of the cantrev belonged to the whole family in common, and was partitioned among the males to the fourth generation, but no one could transfer or sell his rights without first obtaining the sanction of the whole family. The son was not compelled to wait till his father died ere he obtained land ; he received his portion as soon as he attained to man's estate.

The division into cantrevs and cwmmwds is of very early origin, and the limits of each appear to have been well-established and generally recognised in the tenth century, when Howel Dda compiled his famous code of laws and collected his catalogue of Welsh customs.

The Cantrev of Meirionydd was first known as Cantrev Orddwy, that is, the Hundred of the Ordovices. When, however, the Gaels were driven out by the men of Cunedda, it was called the territory of Meirion, after the northern chieftain's grandson. The limits of this hundred were the tidal estuary of the Dovey on the south, and the main water-parting of the county on the north. Its religious and ecclesiastical centre was at Towyn, in the notable church founded by St Cadvan in the sixth century, the mother church of the hundred. Its commotes of Ystumaner and Talybont were separated by the river Dysynni. These names are suggestive of two ancient strongholds of the lords of Meirionydd. Ystumaner has its fortified rock of Castell-y-Bere which ultimately was crowned with a medieval castle. Talybont near the village of Llanegryn has its ancient primitive mound, while the annals of the county tell us of a castle of Cymmer. We have thus the temporal strongholds and the spiritual retreat within the compass of the territory.

The Cantrev of Penllyn is encompassed on all sides by high and rugged mountains, with Llyn Tegid at its

centre. The old home of Llywarch Hen, the veteran
warrior and princely poet of the seventh century, stands
here. He held his court on the mound near Llanfor
church which still bears his name. King Arthur of
legend and romance is reputed to have lived on the lake-
side at Caergai, which appears to have given the domain
its name of Penllyn, although in later times Y Bala
("The Outlet") was the site of the chief stronghold of
the cantrev. Arwystli, the cantrev of the head waters
of the Severn, is not now a part of our county.

The county is now divided into the five hundreds
of Ardudwy, Penllyn, Ystumaner, Talybont, and
Edeyrnion. By the Act of 1536, Arwystli was annexed
to Montgomeryshire, whilst the commotes of Edeyrnion
and Glyndyfrdwy were taken from Powys and annexed to
Meirionydd, and the independent and lawless lordship
of Mawddwy was similarly attached.

In English law every hundred was divided into
townships, or, as they are now called, parishes. Every
township was privileged to have its own local assembly
or parliament, where every freeman had a right to appear,
and in which they appointed their own officers to carry
out the laws. The whole country was divided into
parishes as early as the reign of Edward III. In addition
to the courts of the shire and hundred, there were courts
of the manor presided over by the "Lord of the Manor."
These courts were known as court-leets, where the lord
met his tenants, and arranged all matters which pertained
to the manor, such as the holding of fairs and markets,
and the privilege of common rights. This court continues

to be held in most counties once a year, but it is kept up mainly because of its antiquity.

The manors of olden times may be said to correspond in a measure with the ecclesiastical parishes, of which we have 43 situated wholly or in part within the ancient geographical county of Merioneth. The parishes vary very much in size, number of houses, and population. Some are large and some are very small. The two largest parishes in the county are Llanfor and Trawsfynydd with over 30,000 acres in each ; the three next in point of size are Dolgelly, Llanuwchllyn, and Llanycil with over 20,000 acres in each. The smallest parish of the county is Llansantffraid-Glyn-Dyfrdwy, which has only 670 acres.

In the administrative county there are 39 civil parishes, which are grouped together for the care of the poor into five Poor Law Unions, over each of which there is a Board of Guardians. The workhouses are at Bala, Corwen, Dolgelly, and Penrhyndeudraeth, where the destitute and incapable are given employment and cared for.

The Act of 1888 brought into existence the County Councils, whose powers cover the county as a whole. The County Council levies rates and borrows money for public works subject to the sanction of the Local Government Board. It is responsible for the whole administrative business of the county, such as carrying into effect the laws passed by Parliament, keeping roads and bridges in good repair, managing lunatic asylums and reformatories, and exercising control over Education, both elementary and secondary ; while in conjunction with the Quarter Sessions it manages the police affairs, and appoints coroners and officers to look after the health of the community. The Merionethshire County Council, which meets in rotation at Dolgelly, Bala and Festiniog, consists of 57 members, of whom 43, called councillors, are elected every three years by ratepayers of the county, while 14, called aldermen, are elected or co-opted by the councillors for a period of six years.

In 1894 was passed the Parish and District Councils Act, which confers upon parishioners a good deal of power in the management of local affairs. There are 31 Parish Councils in this county, and six Town and Urban District Councils located at Bala, Barmouth, Dolgelly, Festiniog, Mallwyd, and Towyn, together with five Rural District Councils at Deudraeth, Dolgelly, Edeyrnion, Penllyn, and Pennal.

For the administration of justice the county is in the North Wales and Chester Judicial Circuit, the Assizes being held at Dolgelly twice a year. The county is very free from crime and the judges on circuit are often presented with white gloves. It has one court of Quarter Sessions held at Bala and Dolgelly four times a year, while Petty Sessions presided over by local justices of the peace are held at six centres, namely at Dolgelly for the petty sessional division of Talybont; at Bala for the division of Penllyn; at Towyn for the division of Ystumaner; at Corwen for the division of Edeyrnion; at Blaenau Festiniog for the division of Ardudwy-uwch-Artro; and at Barmouth for the division of Ardudwy-is-Artro. The police force consists of a Chief Constable, one superintendent, two inspectors, five sergeants, and 25 constables, making a force of 34 men, with its head office at Dolgelly.

The chief person in the county in an official capacity is the Lord Lieutenant, who in virtue of his office occupies a similar position to the ealdorman of former times. He is the personal representative of the Sovereign,

Dolgelly

who appoints him, and is either a nobleman or a large land-owner. He remains in office for life.

The next official of the greatest importance is the High Sheriff, who corresponds to the ancient shire-reeve of earlier times. His appointment is an annual one, made on "the morrow of St Martin's Day," that is, November 12th, by the King. The Sheriff holds office for a twelve-month only, and it entails upon him very great expense. His main duty is legal, and he usually appoints an Under Sheriff, who may be a solicitor or a person well-versed in the law. The Sheriff has to make all the necessary arrangements for his Majesty's Justices of Assize when they visit the county to try the cases sent to them from the Quarter Sessions.

The parliamentary representation of Merionethshire is limited to one member. The county has no parliamentary nor municipal borough.

In ecclesiastical matters the county comprises 43 parishes, and is included in part in the diocese of Bangor, and in part in that of St Asaph.

The market towns are Dolgelly and Corwen.

23. The Roll of Honour of the County.

Of the Merionethshire men who have reached high positions as statesmen and lawyers we have Baron Lewis Owen, a native of Dolgelly. He occupied the important office of Vice-Chamberlain and Baron of the Exchequer of North Wales in the reigns of Henry VIII, Edward VI, and Mary. He was member for the county in the parliaments of 1547, 1552, and 1554, and was High Sheriff in 1546, and in 1555—the year he was murdered. Colonel John Jones of Maesygarnedd, the Regicide, played an important part in the great civil war of the seventeenth century. He held high office under Commonwealth rule, being one of the Lord Justices of Ireland. He married a

sister of Oliver Cromwell, and was made by the Protector
a member of the House of Lords.   Baron Richard
Richards of Caerynwch became Baron of the Exchequer
in 1814, and Lord Chief Baron three years later.   His
eldest son Richard was one of the Masters in Ordinary

Thomas Edward Ellis

of the High Court of Chancery, and M.P. for the county
at the time of his father's death.   Sir John Williams,
Solicitor-General in the Parliament of 1830, became a
Baron of the Exchequer and won high repute as a criminal
judge.   Thomas Edward Ellis of Cynlas, the son of a

humble farmer, by sheer ability and force of character became one of the most potent influences in the political life of Wales in the closing years of the last century. He represented his native county in Parliament for some years, and at the time of his death was Chief Whip in Lord Rosebery's government.

During the continental wars of the eighteenth century General Henry Lloyd, the son of a clergyman, greatly distinguished himself. He served in the Russo-Turkish war of 1774, won great renown at the battle of Silistria, and later was entrusted with the command of thirty thousand men in the war with Sweden. He wrote several works on military matters, in addition to his well-known *History of the Seven Years' War*.

In the dark days of the seventeenth century no name stands out more clearly than that of Morgan Llwyd of Gwynedd, of the family of Cynfael. He laboured with great zeal on behalf of Free Church principles. His famous *Llyfr y Tri Aderyn* is a Welsh classic. Another eminent Nonconformist preacher of this period was Hugh Owen of Bronyclydwr, the grandson of John Lewis Owen of Llwyn near Dolgelly, M.P. for the county. The great theologian and divine of the seventeenth century, Dr John Owen, was of the same Merionethshire stock as Hugh Owen. He was an attached friend to Cromwell, and was Vice-Chancellor of the University of Oxford.

The county is justly proud of Thomas Charles, who, though of Caermarthenshire birth, did his life-work at Bala. He was the founder of the Welsh Sabbath School movement, an organisation which has made the Welsh the best versed of any people in the scriptures. The world-famous "British and Foreign Bible Society" was established by his influence and inspiration. His granddaughter was married to Dr Lewis Edwards, who together with his brother-in-law, David Charles, founded the Bala Calvinistic Methodist College in 1837. His son, Dr Thomas Charles Edwards, the first Principal of Aberystwyth University College, was born at Bala. The influence of father and son upon the educational life of Wales cannot be estimated. David Charles, after five years co-operation with his brother-in-law, was invited to start Trevecca College upon the same lines as Bala, and acted as its principal for 20 years. Simon Lloyd, of Plasyndre, Bala, worked with Thomas Charles in the early Methodist movement, and after the death of Charles edited two volumes of *Y Drysorfa*.

In the ranks of great churchmen there stand forth the names of Ellis Wynne of Lasynys, "Y Bardd Cwcs," as he is familiarly called, and Edmund Prys, famous for his metrical version of the Psalms. Dr William Lloyd of Llangower was chaplain to Charles II. He was made Bishop of Llandaff in 1675, was translated to Peterborough in 1679, and thence to Norwich in 1685. Dr Humphrey Humphreys, a native of Penrhyndeudraeth, became Bishop of Bangor in 1689, and was translated to Hereford in 1701. Dr John Thomas, a native of Dolgelly, was the son of very poor parents. He showed early genius and was educated at the Merchant Taylors' School and Cambridge University by his father's employer. He went abroad and became chaplain to the English factory at Hamburg. In 1743 he was made Bishop of St Asaph, but before being consecrated he was translated to Lincoln. In 1761 he was again translated, this time to Salisbury. David Lloyd, born at Trawsfynydd in 1635, became reader of the Chapter-house. He was canon of St Asaph and chaplain to the bishop, and author of several interesting works, his best known being *Statesmen and Favourites of England since the Reformation*. John Ellis, D.D., Archdeacon of Meirionydd, was a learned antiquary and collaborated with Browne Willis in collecting material for the latter's survey of the diocese of Bangor. Griffith Hughes, a native of Towyn, became rector of St Lucie's, Barbados. He is best known as a naturalist; his *Natural History of Barbados* is a valuable work, for which he was elected a Fellow of the Royal Society in 1750. William Wynne, of Maesyneuadd, was an excellent Welsh poet. His poems are refined and classical. Several of these may be seen in a little work published in 1759, and entitled *Dewisol Ganiadau yr Oes hon*. Evan Lloyd, one of the Garrick circle of literary men, was born at Frondderw near Bala, and became vicar of Llanfair Dyffryn Clwyd. He possessed great poetic abilities. His best known poems are *The Power of the Pen*, *The Curate*, and *The Methodist*. His satirical allusion to a neighbouring squire, in *The Methodist*, brought him into trouble, and he was imprisoned in the King's Bench at the same period as the famous John Wilkes, who wrote the verses now inscribed on his tombstone in Llanycil church. John Jones, better known by his bardic name of "Ioan Tegid," a native of Bala, became Vicar of Nevern and prebendary of St David's Cathedral. His scholarly knowledge of his native tongue made him one of the greatest authorities of his time in Welsh orthography.

Conspicuous among Nonconformist preachers we have Robert Thomas, "Ap Vychan," a native of Llanuwchllyn, who was Professor of Theology at Bala. Roger Edwards a native of Bala was the author of *Y Tri Brawd* a religious novel, and Editor of *Y Drysorfa* and *Traethodydd*. Richard Humphreys of Dyffryn was one of the most famous of Methodist divines, and his grandson Richard Humphreys Morgan is noteworthy as the adapter of Pitman's system of shorthand to the Welsh language. Finally, Evan Jones, "Ieuan Gwynedd," will live long in the affection of his countrymen for his patriotic labours.

The Welsh nation will ever be under a deep debt of obligation to Robert Vaughan of Hengwrt, the great scholar and antiquary, born in 1592, for his unrivalled private collection of Welsh MSS. already referred to. The copies transcribed by him were made more valuable by his own scholarly notes and copious annotations. He died at Hengwrt in 1666. Rowland Vaughan of Caergai, who was High Sheriff of the county in 1643, belonged to an equally ancient branch of the Vaughan family as did the squire of Hengwrt. He did much to improve the social condition of his poorer countrymen, translating many excellent works into the vernacular, which he published at his own cost and distributed among the poor. Vaughan suffered for his devotion to King Charles by having his mansion at Caergai burnt to the ground in 1645 by the Parliamentary forces. Dr William Owen Pughe, the lexicographer, was a native of Llanfihangel-y-Pennant, and his son Aneurin Owen, who held important offices under the Government, was the author of the well-known work entitled *The Ancient Laws and Institutes of Wales*. The county is deservedly proud of John

Tyn-y-Bryn
(*Birthplace of William Owen Pughe*)

Griffiths, "Y Gohebydd," who was a native of Barmouth. He was the London correspondent of the *Baner ac Amserau Cymru*, and aided Sir Hugh Owen in the establishment of British schools in all parts of Wales.

Among poets and bards we have Llywelyn Goch ap Meurig of Nannau, who flourished in the fourteenth

century. Six of his poems are printed in the *Myvyrian
Archaeology*, and his elegy was written by Iolo Goch, the
bard of Owain Glyndwr. Sion Phylip of Ardudwy and
William his brother were eminent poets. The former
died in 1620, and the latter in 1669. William suffered
cruel treatment on account of his Royalist leanings, and
his property was confiscated by the Commonwealth.
David Richards, better known as " Dafydd Ionawr," the
author of *Cywydd y Drindod*, was a native of Towyn.
John Phillips, " Tegidon," was born at Llanycil, and
Rice Jones, the compiler of *Gorchestion y Beirdd*, was a
native of Blaenau Festiniog. Hugh Jones, "Maesglasau,"
Hugh Derfel Hughes, David Ellis, compiler of *Y Piser
Hir*, and Edward Hughes, "Y Dryw," as well as many
others, swell the roll. Among the younger men there
is one who will ever hold an honoured place—Robert
Owen of Tai Croesion, who died when only 27 in Aus-
tralia in 1885. His poems are among the most pathetic
and touching in the language.

Among musicians there are Edward Jones, " Bardd y
Brenin," a native of Llandderfel, who filled the post of
bard and harpist to the Prince of Wales in 1774, and
Edward Stephens, " Tanymarian," a native of Fes-
tiniog, and the author of the oratorio, *The Storm of
Tiberias*.

Among distinguished men in the medical profession
we have Gruffydd Owen, a native of Dolgelly, who
emigrated to America in the early years of the eighteenth
century, and was the first physician of the newly-formed
State of Pennsylvania. Professor Alfred William Hughes

of Corris ranks among the most famous of Welsh surgeons. He was for a time at the head of the School of Anatomy at the University College of Cardiff, which he relinquished to become professor of Anatomy at the London University. During the last great war in South Africa, he went out as a surgeon to organise the Welsh Hospital for the troops, and died there of fever. His memory has been perpetuated by the erection of a national memorial at his native place.

Merionethshire may well be proud of Edward Edwards, the marine zoologist and inventor of aquaria for the preservation of fish. He was a native of Corwen, and in 1864 was led to study the habits of fish in their native element among the fissures and rocks of the Menai Straits; and step by step he perfected an invention for keeping fish in health in confinement. The principle of his tank has been adopted in all aquaria in our country as well as on the Continent and in America.

24. THE CHIEF TOWNS AND VILLAGES OF MERIONETHSHIRE.

(The figures in brackets after each name give the population of the parish in 1911, from the official returns, and those at the end of each paragraph are references to the pages in the text.)

Aberdovey (1466) is a small seaport on the Dovey estuary. It exports slates from Corris and Aberllefenni to the extent of nearly 5000 tons annually. It has considerable coasting trade and a regular communication with Ireland. Much fishing is done in the estuary. There is a station on the Cambrian Railway. Direct communication is made with Borth on the opposite coast by means of a ferry which serves at high tide. Its golf links enjoy a national reputation, and the mildness and salubrity of its climate render it a favourite resort at all seasons of the year. (pp. 8, 45, 51, 65, 79, 80, 82, 90, 135, 136.)

Abergynolwyn is a village seven miles inland north-east of Towyn, with which communication is maintained by a little narrow-gauge railway. The main occupation of the people is slate-quarrying. The village lies partly in the parish of Tal-y-llyn and partly in that of Llanfihangel-y-Pennant. (pp. 35, 71, 135, 138.)

Aberllefenni is a village in the Corris quarry district. It stands on a feeder of the Dulas river, and is charmingly situated in a secluded glen, being sheltered on the north by Mynydd Ceiswyn, an offshoot of the Cader Idris ridge. It is reached by the Corris narrow-gauge railway from Machynlleth, which terminates here. (pp. 35, 71, 138.)

Arthog is a small village near the Mawddach mouth, and only a very short distance from Barmouth Junction. The Dolgelly branch of the Cambrian Railway has a station here. Near the village is the fine mansion of Garthyngharad, formerly the residence of the Wyatts. (pp. 46, 71, 129, 135, 136.)

Bala (1537) a small town and the head of a Poor Law Union district is situated at the north-east end of Llyn Tegid. The Great Western Railway has a station here on its Ruabon and Festiniog branch line. It is a seat of Petty Sessions, and the Quarter Sessions are held here in April and October. The place is a great resort of anglers and tourists. Flannel manufacture and brewing are the chief industries; in former times it was the centre of a great trade in stockings and socks. The Calvinistic Methodist Theological College stands on rising ground a little outside the town. Until recent years there was an Independent College here also, but this has been transferred to Bangor. In front of Capel Tegid there stands a fine monument to Thomas Charles, the founder of Sunday Schools. It possesses, too, a fine bronze statue of the late T. E. Ellis, M.P. "The Green" an open space near the railway station is celebrated in poetry and song for its great religious assemblies. The "Tomen y Bala" is an ancient artificial mound. The whole neighbourhood is rich in historical and traditional lore. (pp. 18, 23, 34, 36, 42, 54, 78, 127, 136, 143.)

Barmouth (2106), on the northern side of the Mawddach estuary, is a fashionable watering place. By the Welsh it is called Abermaw. The Cambrian Railway has a station here. The line is carried across the broad estuary by a long timber-trestle bridge 800 yards in length, which also has a gangway for foot passengers from Arthog and Barmouth Junction. Its prosperity depends upon the visitors and tourists. It has a fine stretch of sandy beach, and the commanding views from the high ground behind the town are very fine. The town is governed by an Urban District Council. It possesses a Secondary School under the Welsh Intermediate Education Act. (pp. 17, 34, 46, 56, 65, 70, 76, 99, 112, 134, 136, 141, 143.)

Bettws-Gwerfil-Goch (232) is a hamlet and a parish three miles west of Gwyddelwern station on the L. and N. W. Railway, and six miles north-west of Corwen. The district is purely rural. At Bottegir, now a farmhouse, lived the famous Colonel Salisbury, familiarly known as "*Hosannau Gleision*," "Blue Stockings," the sturdy Royalist Governor of Denbigh Castle, who defended it against the Parliamentary forces for fourteen weeks.

Bontddu is a small village beautifully situated half-way between Barmouth and Dolgelly. The Clogau gold mines in the woody dell which divides the village in two, give employment to many hands. (pp. 21, 75, 76, 134.)

Brithdir is a small village three and a half miles to the north-west of Dolgelly, Bontnewydd is its nearest railway station. (p. 52.)

Corris (1079) on the Dulas, a tributary of the Dovey, is a village and a township in Tal-y-llyn parish. It lies six miles to the north-east of Machynlleth. The great majority of the men are engaged in the slate-quarries, where slate of an excellent quality is obtained. (pp. 35, 70.)

Corwen (2856) is a market town and a parish on the Dee, some ten miles to the west of the town of Llangollen. It has a station at the junction of the Llangollen and Bala branch of the G. W. R. with the Chester, Denbigh, and Ruthin branch of the L. and N. W. R. The town is the head of a Poor Law Union and County Court District, and also a seat of Petty Sessions. Slate is quarried in the neighbourhood and there are a few small flannel factories which give employment to many hands. It is a quiet old-fashioned town and is much frequented by anglers.

The church is of ancient date and contains a monument to Iorwerth ap Sulien; in the churchyard there is an eighth century cross. Rûg mansion near by, a seat of the Wynne family, contains a knife and a dagger said to have been the property of Owain Glyndwr. (pp. 36, 42, 54, 70, 92, 98, 129, 136, 138, 142, 143, 145.)

Corwen

Dinas Mawddwy at the terminus ot the Mawddwy narrow-gauge railway from Cemmaes Road on the Cambrian Railway was at one time an important corporate town, but is now a decayed village. It is ten miles south-east of Dolgelly. Formerly it was one of the five independent lordships of Wales, and was not united to the county of Merioneth until the reign of Henry VIII, when it bore a very ill name. Some of its ruffians cruelly murdered the Baron Lewis Owen because he had condemned eighty of their confrères to suffer the extreme penalty of the law for various crimes in 1554. The spot in the woods of Mawddwy where he was murdered when returning from the Montgomery Sessions in 1555 is still known as Llydiart-y-Barwn. A court leet is still held there twice a year. Some slate-quarrying is carried on, and it is famous for its wild and romantic scenery. (pp. 23, 36, 76, 77, 81, 96, 97, 138.)

Dolgelly (2160) is a market and county town on the Wnion. It is the head of a County Court division. The Cambrian and G. W. Railways run into the town, and it is a convenient centre for tourists and anglers. It has a manufactory of flannels and coarse woollen cloths; tanning and currying are also carried on here. Two weekly papers are printed and published in the town—*Y Goleuad* and *Y Dydd*. A path leads from the town to the summit of Cader Idris. The surrounding scenery is extremely varied and beautiful. An ancient house called Cwrt Plas-yn-dre stood here till lately, associated traditionally with Owain Glyndwr, an interesting specimen of sixteenth century architecture. In 1404 Glyndwr dated his letter from the town, when he entered into an alliance with the King of France against Henry IV. The Dolgelly Grammar School is an anciently endowed foundation which has turned out several men of note. Dr Williams's Endowed School is a famous school for girls, and the County School under the Welsh Intermediate Education Act gives abundant opportunities for secondary education. (pp. 18, 25, 27, 35, 37, 42, 70, 77, 92, 93, 100, 125, 130, 131, 134, 135, 136, 142, 143, 144, 145.)

Dyffryn is a straggling village near the coast about five miles to the north of Barmouth. Passengers by train alight at the little station when they intend to proceed to Drws Ardudwy and the Roman Steps. (pp. 48, 136, 150.)

Festiniog (9682) is a large slate-quarrying area, having many industrial villages, of which the chief are Llan, Blaenau, Conglywal, Rhiw, Manod, and Tan-y-Grisiau. It has an Urban District Council. The G. W. R. and L. and N. W. R. have communications with the place, as well as the Toy Railway from Portmadoc. The place has many chapels, churches, elementary and secondary schools. (pp. 27, 35, 70, 72, 82, 136, 143.)

Gwyddelwern (711) is a village and a parish near the borders of Denbighshire, and about two and a half miles north of Corwen. It has a station on the Corwen and Denbigh branch of the L. and N. W. Railway.

Harlech is an ancient little place, ten miles to the north of Barmouth. It was at one time the county town, and enjoys the unique privilege of being a free borough, whatever that may mean, since the reign of Edward I. The Cambrian Railway has a station here. Its renowned castle gives the place its significance, for it has played important parts in the history of our land from very early times. The Harlech golf links are now considered among the best in Wales. (pp. 34, 48, 76, 80, 95, 96, 101, 102, 104, 118, 120, 121, 136.)

Llanbedr (320) is a pretty little village and a parish on the river Artro, two miles south of Harlech. It has a station on the Cambrian Railway, and is one of the best fishing stations in the county. It is a centre for visitors exploring the Ardudwy country. (pp. 21, 48, 76, 80, 110, 131.)

Llandderfel (785) is a parish and a township on the river Dee, four miles to the east of Bala. It has a station on the G. W. R. The Dee in its meanderings by the village proceeds through charming scenery. The church of Derfel Gadarn is an interesting structure, containing some curious relics, among which is a wooden crosier. (pp. 18, 43, 52, 113, 130, 136.)

Llandrillo (591) a village and a parish five miles to the south-west of Corwen. It has a station on the G. W. R. Slate is quarried in the neighbourhood. The village is the starting point for visiting the waterfall known as Pistyll Rhaiadr, and for the ascent of the Berwyns. (pp. 18, 109, 114, 136.)

Llanegryn (560) is a village and a parish along the southern bank of the Dysynni river. The village is situated some two and a half miles from the coast. The township of Peniarth is comprised in the confines of the parish. (pp. 23, 111, 113, 116, 140.)

Llanelltyd (450) is a parish and a small village on the upper Mawddach two miles distant from Dolgelly. Cymmer Abbey and the Hengwrt mansion are in the neighbourhood. (pp. 20, 82, 109.)

Llanfachreth (711) is the name of a parish, and a village situated three and a half miles to the north-east of Dolgelly. It is almost at the source of the Mawddach. It contains the famous old mansion of Nannau. At one time it was noted for its mineral resources. (pp. 17, 76, 116.)

Llanfihangel-y-Pennant (457) is a parish, and a small village situated about seven miles to the north-east of Towyn. The village is on the Dysynni. The church is ancient and has a rare specimen of a leper window. Castell-y-Bere lies in the parish. (pp. 23, 109, 111.)

Llanfrothen (861) is the name of a parish in the reaches of the Traeth Mawr. The village is about one mile from Penrhyndeudraeth.

Llangar (572) is a parish on the Dee; the township is at the influx of the Alwen into the Dee. It is one and a half miles to the south-west of Corwen. (pp. 110, 114.)

Llansantffraid-Glyn-Dyfrdwy (170) is a small parish and a village situated on the Dee, two miles to the east of Corwen. Its nearest station is Carrog on the G.W.R. The chief residence is Rhagatt. (pp. 19, 142.)

Llanuwchllyn (1007) is a large parish at the western end of Bala Lake. It contains many old mansions and places of great interest connected with the history and antiquities of the county. The village has a station on the G. W. R. (pp. 18, 76, 109, 136, 142.)

Llanycil (904) a hamlet and a parish on the shores of Bala Lake. In its churchyard there lie buried many famous men, among whom are Thomas Charles, Dr Lewis Edwards, Professor John Peters and Dr Hugh Williams. (pp. 109, 142.)

Llan-y-Mawddwy (323) is a parish at the head of the upper waters of the Dovey, under the Aran Mawddwy mountain. The village is peopled in the main by slate-quarrying folk. A retired spot called Gwely Tydecho close to the roadside is said to be the retreat of the saint of that name. (pp. 23, 131.)

Llwyngwril is a pretty little village in Llangelynin parish situated near the coast, six and a half miles to the north of Towyn. It has a station on the Cambrian Railway. There are many ancient remains in the neighbourhood, the chief being Castell y Gaer close to the village. It has an interesting old burial ground belonging to the Quakers; the date 1646 is inscribed on the entrance gate. The village is becoming a favourite resort of visitors during the summer months. Just two miles to the north is Y Friog, a small village contiguous to the new health resort of Fairbourne. (pp. 46, 129, 135.)

Maentwrog (652) is a charming place, and a parish in the Dwyryd valley. The Sarn Helen of the Romans traverses the parish. The village is ten miles to the north-east of Harlech. It has taken its name from a monumental stone of great size erected to the memory of Twrog, son of Cadvan, a sixth century saint. Archdeacon Edmund Prys was rector of the parish, and his body has been buried in the church. Here he translated the Psalms into the Metrical Version. (pp. 22, 35, 104.)

Mallwyd (757) is a village and a parish lying partly in this county and partly in Montgomeryshire. The village stands on the river Dovey in the midst of beautiful mountain scenery. It is a favourite resort of anglers and artists. The church is an ancient building made famous by its vicar Dr John Davies, the eminent scholar and antiquary. (pp. 109, 143.)

Pennal (430) a village and a parish on the Sarn Helen lies four miles to the west of Machynlleth. By some authorities this place is identified with the Maglona of the Romans. In the grounds of Talgarth Park stands a huge mound or *tomen* which has yielded many Roman coins to excavators. (pp. 103, 105.)

Penrhyndeudraeth (1988) is a pleasantly situated village at the head of the Traeth Bach. It is served by the Cambrian and the Festiniog Toy Railway. The men who live here are mostly engaged in the quarries of Festiniog. On the northern side of the Traeth standing in its own lovely grounds is the modern castellated mansion of Deudraeth, the seat of the Lord Lieutenant of the County. (pp. 49, 134, 136, 142, 143.)

Towyn (3929) is a pleasant seaside town with fine sands and promenade. The Cambrian railway passes through the place and it holds communication with the slate quarrying district of Abergynolwyn by means of a narrow-gauge railway. Its church is of ancient foundation. The town is provided with a Secondary School under the Welsh Intermediate Education Act. (pp. 45, 56, 65, 67, 76, 79, 80, 81, 109, 110, 113, 128, 135, 138, 140, 143.)

Trawsfynydd is a quarrymen's village situated in an elevated area on the Afon Prysor, about five miles south of Festiniog, and eleven miles to the north of Dolgelly. The Bala and Festiniog section of the G. W. R. passes near to the village, and the Sarn Helen leads through it from the south on its way to Tomen-y-Mur, some two miles further north. The neighbourhood has many ancient remains in the form of encampments, tumuli, stone circles, etc. Great efforts at gold-mining have been made at various times in the neighbourhood. The lakes of the parish abound in fish and are famous for excellent sport. (pp. 23, 27, 89, 102, 109, 142.)

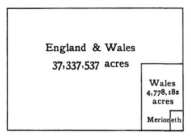

Fig. 1.   Area of Merionethshire (422,372 acres) compared
with that of England and of Wales

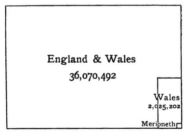

Fig. 2.   Population of Merionethshire (45,565) compared
with that of England and of Wales in 1911

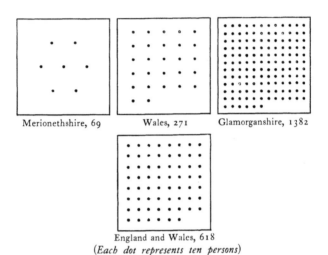

Merionethshire, 69   Wales, 271   Glamorganshire, 1382

England and Wales, 618

(Each dot represents ten persons)

Fig. 3. Comparative Density of Population per square mile in Merionethshire, Wales, England and Wales, and Glamorganshire in 1911

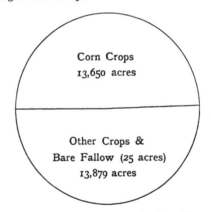

Corn Crops
13,650 acres

Other Crops &
Bare Fallow (25 acres)
13,879 acres

Fig. 4. Proportionate area under Corn Crops in Merionethshire in 1912

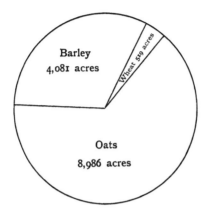

Fig. 5.   Proportionate areas of chief Cereals in
Merionethshire in 1912

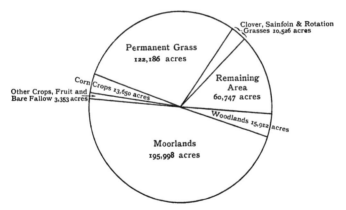

Fig. 6.   Proportionate areas of land in Merionethshire
in 1912

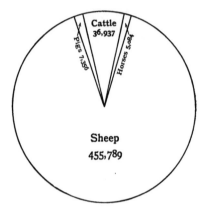

Fig. 7. Proportionate numbers of Live-stock in
Merionethshire in 1912